Structural Time S
and Modeller and Predictor

STAMP™ 8.2

Structural Time Series Analyser
and Modeller and Predictor

STAMP™ 8.2

Siem Jan Koopman
Andrew C. Harvey
Jurgen A. Doornik
Neil Shephard

Published by Timberlake Consultants Ltd
www.timberlake.co.uk
www.timberlake-consultancy.com
www.oxmetrics.net

Structural Time Series Analyser and Modeller and Predictor - STAMP ™8.2
Copyright © 2009 Siem Jan Koopman, Andrew C. Harvey,
Jurgen A. Doornik, and Neil Shephard
First published by Timberlake Consultants in 1999
Revised 2001, 2006, 2009

All rights reserved. No part of this work, which is copyright, may be reproduced or used in any form or by any means – graphic, electronic, or mechanical, including photocopying, record taping, or information storage and retrieval systems – without the written consent of the Publisher, except in accordance with the provisions of the Copyright Designs and Patents Act 1988.

Whilst the Publisher has taken all reasonable care in the preparation of this book, the Publisher makes no representation, express or implied, with regard to the accuracy of the information contained in this book and cannot accept any legal responsibility or liability for any errors or omissions from the book, or from the software, or the consequences thereof.

British Library Cataloguing-in-Publication Data
A catalogue record of this book is available from the British Library

Library of Congress Cataloguing-in-Publication Data
A catalogue record of this book is available from the Library of Congress
Siem Jan Koopman, Andrew C. Harvey, Jurgen A. Doornik, and Neil Shephard
p. cm. – (Structural Time Series Analyser and Modeller and Predictor – STAMP™ 8.2)

ISBN 978-0-9557076-2-9

Published by
Timberlake Consultants Ltd
Unit B3, Broomsleigh Business Park
London SE26 5BN, UK
http://www.timberlake.co.uk

842 Greenwich Lane,
Union, NJ 07083-7905, U.S.A.
http://www.timberlake-consultancy.com

Trademark notice
All Companies and products referred to in this book are either trademarks or registered trademarks of their associated Companies.

Contents

List of Figures   xv

List of Tables   xviii

Preface   xix

The Authors   xxi

I   Prologue   1

1   Introduction   3
- 1.1   Overview of the STAMP book   3
- 1.2   General information   4
- 1.3   Features introduced in STAMP 7   5
- 1.4   New in STAMP 8   5
- 1.5   Developments in STAMP 8.20   8
- 1.6   The special features of STAMP   8
- 1.7   Basics of the program   10
  - 1.7.1   Data storage and input   10
  - 1.7.2   Menus and dialogs   11
  - 1.7.3   Help system   12
  - 1.7.4   Results storage   12
- 1.8   Using STAMP documentation   12
- 1.9   Citation   13
- 1.10   World Wide Web   13
- 1.11   Tutorial data sets   14
  - 1.11.1   ENERGY: energy consumption in the UK   14
  - 1.11.2   EXCH: daily exchange rates for the US dollar   15
  - 1.11.3   INTEREST: short- and long-term interest rates   16
  - 1.11.4   MINKMUSK: minks and muskrats in Canada   16
  - 1.11.5   NILE: level of the Nile   16
  - 1.11.6   RAINBRAZ: rainfall in north-east Brazil   16

	1.11.7	SEATBELT and SEATBQ: road casualties in Great Britain and the 1983 seat belt law	16
	1.11.8	SPIRIT: consumption of spirits in the UK	17
	1.11.9	TELEPHON: telephone calls to three different countries	17
	1.11.10	UKCYP: consumption, income and prices in the UK . .	17
	1.11.11	USYCIMP: US macroeconomic time series	17
	1.11.12	USmacro07: more recent US macroeconomic time series	18
1.12	Data sets used in exercises		18
1.13	STAMP and PcGive		19

2 Getting Started — 20
- 2.1 Starting STAMP 20
- 2.2 Loading and viewing the tutorial data set 21
- 2.3 OxMetrics graphics 23
  - 2.3.1 A first graph 23
  - 2.3.2 Graph saving and printing 25
- 2.4 Data transformations 25

II Tutorials on Structural Time Series Modelling — 29

3 Introduction to Univariate Modelling — 31
- 3.1 Model formulation 31
- 3.2 Evaluating and testing the model 38
  - 3.2.1 More written output 38
  - 3.2.2 Components graphics 39
  - 3.2.3 Weight functions 40
  - 3.2.4 Residuals graphics 42
  - 3.2.5 Prediction graphics 44
  - 3.2.6 Forecasting 47
- 3.3 Exercises 47

4 Tutorial on Components — 49
- 4.1 Selection of components 49
- 4.2 Trend 50
  - 4.2.1 Local level model 50
  - 4.2.2 Statistical analysis of the local level model 52
  - 4.2.3 Local linear trend and smooth trend 53
  - 4.2.4 Statistical specification of the local linear trend model . . 55
  - 4.2.5 Specification of the trend 56
- 4.3 Seasonal 56
  - 4.3.1 Specifying and testing the seasonal component 57

	4.3.2	Seasonal adjustment	59
4.4	Cycle		60
	4.4.1	A simple cycle plus noise model	60
	4.4.2	Statistical specification	63
	4.4.3	Higher order cycles	63
	4.4.4	Trend plus cycle	64
	4.4.5	Multiple cycles	64
4.5	Autoregression		66
4.6	Exercises		66

5 Tutorial on Interventions and Explanatory Variables 68
- 5.1 Interventions 68
  - 5.1.1 Modelling 69
  - 5.1.2 Detection using auxiliary residuals 70
  - 5.1.3 Automatic outlier and break detection 72
  - 5.1.4 Specification of more complex interventions 75
- 5.2 Explanatory variables 77
  - 5.2.1 Stochastic trend component 78
  - 5.2.2 Outliers and structural breaks 82
  - 5.2.3 Lags and differences 82
- 5.3 Forecasting 83
  - 5.3.1 Incremental change 84
  - 5.3.2 Manual input 85
  - 5.3.3 Using models to forecast the explanatory variables . . . 86
  - 5.3.4 Interventions 86
- 5.4 Statistical features of the models 86
- 5.5 Exercises 87

6 Tutorial on Multivariate Models 88
- 6.1 SUTSE models 88
- 6.2 Cycles 90
- 6.3 Autoregression 92
- 6.4 Common factors and cointegration 93
  - 6.4.1 Statistical specification of common levels 94
  - 6.4.2 An example of common levels: road casualties in Britain 95
  - 6.4.3 Balanced levels 97
  - 6.4.4 Common trends 99
  - 6.4.5 Common seasonals 101
  - 6.4.6 Common cycles 101
  - 6.4.7 Factor rotations 102
- 6.5 Explanatory variables and interventions 102
- 6.6 Assessing the effect of the seat belt law using a control group . . . 103

	6.7	Exercises	104

7 Applications in Macroeconomics and Finance — 106
- 7.1 Univariate trend-cycle decompositions: GDP — 106
- 7.2 Multivariate trends and cycles: GDP and Investment — 113
- 7.3 Inflation — 114
  - 7.3.1 Expected inflation — 114
  - 7.3.2 Inflation and the output gap — 116
  - 7.3.3 Bivariate modeling — 119
- 7.4 Stochastic volatility — 125
- 7.5 Seasonal adjustment and detrending — 127
  - 7.5.1 Seasonal adjustment — 128
  - 7.5.2 Detrending — 128
- 7.6 Missing values — 130
  - 7.6.1 Some missing observations in the time series — 130
  - 7.6.2 High-frequency trade prices and many missing observations — 132
- 7.7 Exercises — 135

8 Tutorial on Model Building and Testing — 136
- 8.1 Specification of univariate models — 137
  - 8.1.1 Formulate a model — 137
  - 8.1.2 Selection of components — 140
  - 8.1.3 Specify regression coefficients — 142
  - 8.1.4 Selection of interventions — 143
  - 8.1.5 Edit and fix parameter values — 145
- 8.2 Estimate a model — 147
  - 8.2.1 Estimate dialog — 147
  - 8.2.2 Maximum likelihood — 147
  - 8.2.3 Options — 150
  - 8.2.4 Progress — 151
- 8.3 Model evaluation and testing — 151
  - 8.3.1 Estimation report — 153
  - 8.3.2 Diagnostic summary report — 153
  - 8.3.3 More written output — 155
  - 8.3.4 Components graphics — 158
  - 8.3.5 Weight functions — 160
  - 8.3.6 Residuals graphics — 160
  - 8.3.7 Auxiliary residuals graphics — 161
  - 8.3.8 Prediction testing — 162
- 8.4 Forecasting — 164
  - 8.4.1 Without explanatory variables — 164
  - 8.4.2 Interventions — 165

| | | 8.4.3 | Explanatory variables . . . . | 165 |

III Statistical Treatment   167

9 Statistical Treatment of Models   169

- 9.1 Model definitions 169
  - 9.1.1 Univariate time series models 169
  - 9.1.2 Explanatory variables and interventions 171
  - 9.1.3 Multivariate time series models 171
  - 9.1.4 Common factors 172
    - 9.1.4.1 Common levels 172
    - 9.1.4.2 Smooth trends with common slopes 173
    - 9.1.4.3 Common trends: level and slopes 173
    - 9.1.4.4 Common seasonals 174
    - 9.1.4.5 Common cycles 174
  - 9.1.5 Explanatory variables 175
- 9.2 State space form 175
  - 9.2.1 Structural time series models in SSF 177
    - 9.2.1.1 Univariate models in SSF 177
    - 9.2.1.2 Multivariate model in SSF 178
    - 9.2.1.3 Deterministic trend and seasonal components . 179
- 9.3 Kalman filter 179
  - 9.3.1 The augmented Kalman filter 180
  - 9.3.2 The likelihood function 181
    - 9.3.2.1 Univariate models: the concentrated likelihood 182
    - 9.3.2.2 Multivariate models 183
  - 9.3.3 Prediction error variance 183
  - 9.3.4 The final state and regression estimates 183
  - 9.3.5 Filtered components 184
  - 9.3.6 Residuals 184
    - 9.3.6.1 Generalised least squares residuals 184
    - 9.3.6.2 Generalised recursive residuals 185
- 9.4 Disturbance smoother 185
  - 9.4.1 The augmented disturbance smoother 186
  - 9.4.2 The EM algorithm and exact score 186
  - 9.4.3 Smoothed components 187
  - 9.4.4 Auxiliary residuals 188
- 9.5 Forecasting 188
  - 9.5.1 Forecasts of series and components 188
  - 9.5.2 Extrapolative residuals 189
- 9.6 Parameter estimation 189

		9.6.1	The parameters of STAMP	189

- 9.6.1 The parameters of STAMP 189
  - 9.6.1.1 Variance matrices in a multivariate model ... 190
  - 9.6.1.2 Vector autoregressive components 191
- 9.6.2 BFGS Estimation procedure* 191
- 9.6.3 Estimation of univariate models* 193
  - 9.6.3.1 Initial values 193
  - 9.6.3.2 Strategy for setting parameters fixed 194
  - 9.6.3.3 Strategy for determining the concentrated parameter 194
- 9.6.4 Estimation of multivariate models* 194
  - 9.6.4.1 Initial values 195
  - 9.6.4.2 Strategy for fixing parameters 195
- 9.7 Appendix 1: Diffuse distributions* 196
  - 9.7.1 Collapse of the augmented KF 196
  - 9.7.2 Likelihood calculation 197
  - 9.7.3 Recursive regressions 197
- 9.8 Appendix 2: Numerical optimisation* 197
  - 9.8.1 Newton type methods 198
  - 9.8.2 Numerical score vector and diagonal Hessian matrix .. 199

10 Statistical Model Output   200

- 10.1 Output from STAMP 200
  - 10.1.1 Model estimation 200
  - 10.1.2 Selected model and estimation output 201
  - 10.1.3 Summary statistics 201
  - 10.1.4 The sample period 201
  - 10.1.5 Test 202
- 10.2 Parameters 202
  - 10.2.1 Variances and standard deviations 202
  - 10.2.2 Cycle and AR(1) 202
  - 10.2.3 Transformed parameters and standard errors 202
- 10.3 Final state 203
  - 10.3.1 Analysis of state 203
  - 10.3.2 Regression analysis 204
  - 10.3.3 Seasonal tests 204
  - 10.3.4 Cycle tests 205
  - 10.3.5 Data in logs 205
- 10.4 Goodness of fit 205
  - 10.4.1 Prediction error variance 205
  - 10.4.2 Prediction error mean deviation 206
  - 10.4.3 Coefficients of determination 206

		10.4.4	Information criteria : AIC and BIC	207
10.5	Components			207
		10.5.1	Series with components	207
		10.5.2	Detrended	207
		10.5.3	Seasonally adjusted	207
		10.5.4	Individual seasonals	208
		10.5.5	Data in logs	208
10.6	Residuals			208
		10.6.1	Correlogram	208
		10.6.2	Periodogram and spectrum	209
		10.6.3	Cumulative statistics and graphs	209
		10.6.4	Distribution statistics	209
		10.6.5	Heteroskedasticity	210
10.7	Auxiliary residuals			210
10.8	Predictive testing			210
10.9	Forecast			211

A1 STAMP Batch Language — 213

References — 219

Author Index — 223

Subject Index — 225

Figures

1.1	Analysis of gas consumption by final users. Original series and estimated: trend, slope of trend, seasonal and seasonally adjusted series	14
1.2	The estimate of volatility for the DM	15
1.3	Stylised facts of US GNP. Series and estimated trend and cycle. Forecast and forecast confidence intervals	18
2.1	Time-series plot of *ofuGAS* and *ofuGASl*	25
2.2	Time series plot of *absDLPound* and sample ACF of *absDLPound*	28
3.1	Coal consumption by final users. Original series and logarithms .	31
3.2	Components graphics output of model for coal consumption . . .	37
3.3	Components graphics analysis of model for coal consumption, anti-log analysis	41
3.4	Weight graphics analysis for trend of model for coal consumption at period 1973 Q2	42
3.5	Residuals graphics analysis of model for coal consumption	44
3.6	Prediction analysis of model for coal consumption	46
3.7	Forecast analysis of model for coal consumption	47
4.1	Time series of US inflation, from data set USYCIMP	51
4.2	The local level model with forecast of US inflation	52
4.3	The local level model with different estimates of level and irregular components: predicted, filtered and smoothed estimates	54
4.4	Graph and correlogram of US inflation in levels and first differences	54
4.5	The estimated seasonal terms of income in UKCYP	59
4.6	Summaries of the 'RainFort' series	61
4.7	Estimated cycle for 'RainFort' series	62
5.1	Auxiliary residuals for random walk + noise model of the Nile . .	72
5.2	Predictive testing for the Drivers series	78
5.3	Trend and Trend with Explanatory variables for Spirit series . . .	81

xv

LIST OF FIGURES

5.4	Forecasts with Explanatory variables for Spirit series: in upper graph, future Xs are replaced by their realised values in the database; in lower graph, future Xs are extrapolated from their values in period 1930 with an increment of 0.01 for each period	84
6.1	The estimated cycles for minks and muskrats	92
6.2	The estimated trend and autoregressive components for minks and muskrats	93
6.3	Time series plots of estimated slopes for MINKMUSK	99
6.4	Time series plot of estimated trends for MINKMUSK	100
6.5	Smoothed levels of Drivers and Rear Seat Passengers killed and seriously injured in Great Britain, with allowance made for the seat belt law.	105
7.1	US GDP trend-cycle decompositions with a random walk plus fixed drift trend and an AR(2) cycle	109
7.2	Weight and gain functions for the estimated US GDP trend-cycle decomposition based on a smooth trend plus generalised cycle plus irregular model	111
7.3	Weight and gain functions for the estimated US GDP trend-cycle decomposition based on a smooth trend plus generalised cycle plus irregular model (end-of-sample estimates: filtering)	112
7.4	Trends and (fourth-order) cycles in LGDP and LINV from database USmacro07	115
7.5	Predicted, filtered and smoothed estimates of quarterly inflation (from 1990 onwards)	116
7.6	Smoothed components in inflation (INFLcpi in USMACRO07 database)	117
7.7	Smoothed estimates of the cycles obtained from univariate models for LGDP and INFLcpi in USMACRO07 database	117
7.8	Inflation and the output gap as an explanatory variable	118
7.9	Multi-step predictions from end of 1997	119
7.10	Smoothed components from a bivariate model for Inflation and GDP	124
7.11	Estimated $\exp(h_{t|T}/2)$ for the Pound series	127
7.12	Hodrick and Prescott detrending using Algebra in OxMetrics and using local linear trend model in STAMP	129
7.13	Quarterly electricity consumption (other final users) with missing observations	132
7.14	Estimates of components when observations are missing	132
7.15	Weight and gain functions for components when observations are missing	133

7.16	Trade prices (logs) for one day and for the last five minutes (second by second)	134
7.17	Trade price decomposition for one day and for the last five minutes (second by second)	135

Tables

9.1	Some special level and trend specifications	170
9.2	Parameters and transformations	190

Preface

The full name of *STAMP* is *S*tructural *T*ime series *A*nalyser, *M*odeller and *P*redictor. Structural time series models are formulated directly in terms of components of interest. Such models find application in many subjects, including economics, finance, sociology, management science, biology, geography, meteorology and engineering. STAMP bridges the gap between the theory and its application – providing the necessary tool to make interactive structural time series modelling available for empirical work. (Another such tool is *SsfPack*, by Koopman, Shephard and Doornik (1999), which provides more general procedures but with a programmatic interface, see www.ssfpack.com.)

STAMP uses the Kalman filter and related algorithms to fit unobserved component time series models. We are excited to present the new version 8 of STAMP, which provides another big step forward from the previous version. The new version is updated for the new OxMetrics environment and it therefore provides even higher standards in program functionality: by clicking the mouse a few times, everyone is able to start with a full exploratory, statistical or econometric analysis of the time series at hand using the powerful capabilities of the STAMP 8.

Earlier versions of the program were written by Andrew Harvey and Simon Peters, while the data management side was dealt with by Bahram Pesaran. These projects were supported by the Economic and Social Research Council. Version 5 was entirely rewritten in C by the current authors of STAMP. Much of the data management and the graphical interface of STAMP 5 was shared with the PcGive 7 and 8 programs of Jurgen A. Doornik and David F. Hendry. STAMP 6 was the first version of STAMP in which the front-end program GiveWin has been made separate from the econometric module STAMP. Other modules included PcGive^{TM}, TSP^{TM} (by TSP International) and X12Arima for GiveWin (based on X12-ARIMA program of the US Census Bureau). STAMP 7 was a separate module of the OxMetrics 4 program that was developed by Jurgen A. Doornik. The current multivariate program STAMP 8 is developed as a module for the OxMetrics 5 program.

Scientific Word in combination with MikTeX and DVIPS eased developing the documentation for version 7 in LaTeX, further facilitated by the more self-contained nature of STAMP 7 and its in-built help system. We thank Maxine Collett who typed and retyped various sections of the earlier versions of this book. We are also thankful to Tirthankar Chakravarty for his excellent help on editing this book.

We thank the London School of Economics and Political Science for hosting the developments of early versions of STAMP. Furthermore, we are grateful for the comments and encouragements of John Aston, Charles Bos, Jim Durbin, Irma Hindrayanto, Maarten van Kampen, Roy Mendelssohn, Marius Ooms, Jeremy Penzer, Tommaso Proietti, Marco Riani, Thomas Trimbur, Brian Wong and the many old and new users of STAMP who send us suggestions for improvements and bug reports. We hope that you will continue to write us (`s.j.koopman@feweb.vu.nl`) with ideas for improving and extending STAMP (and *report any bugs*!).

The Authors

Siem Jan Koopman is Professor of Econometrics at the VU University Amsterdam and Tinbergen Institute Amsterdam. He is author of the textbooks 'Time Series Analysis by State Space Methods' (with J. Durbin) and 'An Introduction to State Space Time Series Analysis' (with J. Commandeur). He is an Associate Editor of *Journal of Forecasting*.

Andrew C. Harvey is Professor of Econometrics at the University of Cambridge. He is the author of the textbooks 'Time Series Models' and 'Econometric Analysis of Time Series'. He has also published the monograph 'Forecasting, Structural Time Series Models and the Kalman filter'. He is a Fellow of the Econometric Society and the British Academy. He is also an Associate Editor of the *J Time Series Analysis*.

Jurgen A. Doornik is a Research Fellow of Nuffield College, Oxford. He is the main developer of *OxMetrics*, the originator of *Ox*, an object-oriented matrix programming language, and author (with David F. Hendry) of *PcGive* and *PcFiml*.

Neil Shephard is Professor of Economics at the University of Oxford and Research Director of the Oxford-Man Institute of Quantitative Finance. He is a Fellow of the Econometric Society, the British Academy, Nuffield College and is an Associate Editor of *Econometrica*.

Part I

Prologue

Chapter 1

Introduction

1.1 Overview of the STAMP book

(1) *Read this chapter and the next for an introduction to the* STAMP *system.*
This introductory chapter explains how to use the documentation and provides a brief overview of the special features that make STAMP such a powerful tool for time series modelling. Chapter 2 explains how to start STAMP, and gives a brief overview of OxMetrics. The OxMetrics 6 program facilitates data input, text editing, graphics output and much more.

(2) *Read Chapter 3, at the beginning of Part II, for an introduction to how STAMP may be used to model and forecast a time series, and Chapters 4–7 to learn in detail about how structural time series models are applied in practice.*
Chapter 3 provides a simple hands-on introduction to structural time series modelling with STAMP. It is followed by three chapters which teach you how to go about structural time series modelling, and build models using explanatory variables and interventions. The examples provided use a variety of real data sets, from the consumption of spirits in the UK to the water level of the Nile. These data sets are introduced in §1.11. The last chapter in this part of the book deals with applications to economic and financial time series.

(3) Chapters 4–7 require virtually no prior knowledge of structural time series modelling or the theory of time series analysis in general. Hence, the way in which the program may be used is readily accessible to the practitioner. Statistical theory is kept to a minimum with more detailed points being discussed in slanted typescript. Chapter 6 is fully devoted to the analysis and modelling of multivariate models in STAMP.

(4) Chapter 8 systematically describes how univariate and multivariate models are formulated, estimated and evaluated. Users already familiar with structural time series modelling may find it more efficient to consult this chapter first and omit the earlier chapters in Part II.

(5) *Turn to Chapters 9 and 10 in Part III for a description of the statistical output of* STAMP. These chapters define the models, estimation procedures and test statis-

tics available in STAMP. An Appendix lists the commands which are available for the batch use of STAMP.

1.2 General information

STAMP is an interactive menu-driven time series modelling program. Version 7, to which this documentation refers, runs under Microsoft Windows XP. Particular features of the program are its ease of use, edit facilities, flexible data handling, model building, and forecasting capability.

STAMP is designed for modelling time series data using unobserved components. The full name of the program is **S**tructural **T**ime series **A**nalyser, **M**odeller and **P**redictor. The current available module is for the fitting and checking of univariate time series models built out of time varying components of interest such as trends and seasonals. The models can include explanatory variables, interventions, lagged endogenous variables and many other features. The theory of structural time series models is laid out in Harvey (1989), with some additional developments described in Harvey and Shephard (1993) and Harvey (2006). A selection of key papers on unobserved components time series models (statistical theory, estimation and testing, methodology, forecasting) are collected in Harvey and Proietti (2005). An introductory textbook about the structural time series model and the state space approach to time series analysis is Commandeur and Koopman (2007). Basic aspects of structural time series modelling are described from a Bayesian perspective in West and Harrison (1997).

The ease with which structural time series models can be implemented in practice has been demonstrated repeatedly with the previous version of STAMP which was written for the MS-DOS operating system. The new version has been rewritten to enable use of the exciting possibilities offered by the OxMetrics system. The interactive framework of OxMetrics, with extensive use of graphics, offers new scope for modelling. The power of modern machines, coupled with the theoretical and computational advances of state space methods described in Durbin and Koopman (2001) and Koopman, Shephard and Doornik (1999), means that, structural time series modelling is a practical option.

The documentation fully explains the time series methods, the modelling approach and the techniques used, as well as bridging the gap between the time series theory and empirical practice. Detailed tutorials describe the use of the program while teaching the methods. The aim is to provide an operational approach to time series modelling using the most sophisticated yet easy-to-use software available. The context-sensitive on-line help system offers help on the program.

This chapter discusses the special features of STAMP and describes how to use the documentation. The OxMetrics system is described in a separate book, however the next chapter also provides a brief overview of its main features.

1.3 Features introduced in STAMP 7

Many new features were introduced in version 7 of STAMP and they are all kept in the current version 8.20. Some of these features are

- **Missing values** can be treated in this version of STAMP. In all levels of the analysis, missing observations are accounted for automatically. Model-based estimates of the missing values can be produced.
- **Confidence intervals** can be included in the graphs with estimated components. The **predicted**, **filtered** and **smoothed** estimates of the components can be presented simultaneously.
- The observation **weight functions** that are used for the estimation of the unobserved components are given as output. The associating **spectral gain and phase functions** can also be produced.
- Model-based **forecasting** and **backcasting** can be carried out.
- **Time-varying regression coefficients** can be included as part of the model. Three specifications can be selected: (i) random walk, (ii) spline specification (or smooth trend) and (iii) return-to-normality (deviations from a fixed coefficient follow an autoregressive process of order 1).
- Unobserved stationary **autoregressive processes** of orders 1 and 2 can be included in the model together with three stochastic cycle components.
- Higher order **smooth cycles** with band-pass filter properties can be included in the model. Higher order **smooth trends** can also be included and lead effectively to application of Butterworth filters.
- The **trigonometric seasonal component** consists of separate processes for the different seasonal frequencies. These processes can be selected separately in the model. Also different seasonal variances can be attached to the seasonal processes.
- More flexible options for the handling and estimation of **parameters** are provided. The parameters can be treated without transformation.
- Dates for **outliers and breaks** can be stored and remain part of the model once selected.

1.4 New in STAMP 8

Many new features have been introduced in version 8 of STAMP. The most notable are:

- **The multivariate capabilities of the STAMP program are back**. The multivariate structural time series model where the unobserved components become vectors and the disturbance variances become disturbance variance matrices can be considered for the analysis of a set of multiple time series. The number of multivariate options has increased considerably compared to earlier versions of the program:

- *Select components by equation*: Different components can be selected for different equations. This enables the user to analyse time series with different dynamic characteristics jointly. For example, consider two time series where one series may be subject to seasonal dynamics while the other series does not require a seasonal component. The trends of the two time series may move together. STAMP 8 allows the user to select a seasonal component for the first series but not for the second series. This applies to all components in STAMP: trend, seasonal, cycle, autoregressive, irregular, time-varying regressions, etc.
- *Select regressions and interventions by equation*: An option for selecting different explanatory variables and interventions for different equations has been available in STAMP versions 5 and 6. However, the current facility of distributing explanatory variables over different equations has improved and is more flexible.
- *Design a dependence structure for each component*: Multivariate models in STAMP 5 and 6 were limited in their choice of variance matrices: only full variance matrices of different ranks could be considered. A reduced-rank variance matrix implies common features in multiple time series. This option remains in STAMP but the specification has changed slightly. The disturbance variance matrix imposes a dependence structure within the component vector (between the different equations). This dependence can be designed by the user in a simple way and for each component separately. For example, the cycle component in equation 1 can be forced to depend on the cycles in equations 2 and 3 only.
- *In STAMP 8 different variance matrices for different disturbances can be chosen*: The range of variance matrices includes scalar and diagonal matrices, scaled matrices of ones (when applied to the slope component, this implies balanced growth) and one rank plus diagonal matrices. The latter case implies that a vector component can be decomposed into common and idiosyncratic effects. In many applications, these different specifications can be interpreted easily and can be highly interesting.
- *The multivariate options extend to all models introduced in STAMP 7*: This includes the higher-order smooth trend models, the higher-order (bandpass) cycle components and the (vector) autoregressive components of orders 1 and 2.
- *Missing observations*: They can also be handled within multivariate time series models. This allows the interpolation of missing observations through time but also through different time series.
- *Forecasting of multivariate time series made simple*: In particular, STAMP

8 allows the incorporation of available future observations for the explanatory variables in the database. Furthermore, future observations of dependent variables are considered in graphical presentations of forecasts and for the measurement of forecast accuracy (using standard measures such as the root mean squared forecast error (RMSE) and the mean absolute percentage error (MAPE).

- *Estimation of parameters in multivariate time series models is based on exact procedures*: The diffuse initialisation of the Kalman filter is implemented, the exact likelihood function is computed and the score function with respect to variance parameters is computed analytically and fast. This leads to a robust estimation procedure in STAMP 8 and a relatively fast estimation process.
- *The number of graphical output for multivariate models is increased*: STAMP 8 offers an easy handling of the graphical output. An option for graphics output selection for each equation is provided. The powerful tools in OxMetrics 5 to edit graphical output are fully available to STAMP 8 users.

- **Automatic outlier and break detection**: Another major development in STAMP 8 is the implementation of a new automatic detection procedure for outliers and breaks in univariate and multivariate time series models. The following features are available:

  - STAMP 8 is able to propose a set of potential outliers and trend breaks for univariate and multivariate time series. It is a basic but effective two-step procedure based on the auxiliary residuals. First the selected model is estimated and the diagnostics are investigated. Then a first (larger) set of potential outliers and trend breaks are selected from the auxiliary residuals. After re-estimation of the model, only those interventions survive that are sufficiently significant. In the multivariate case, this selection procedure is carried out jointly for each equation in the model.
  - After the automatic selection, the results are reported. All considered outliers and breaks are kept in the intervention dialog and they can be deleted from the model or added to the model in the usual way and implemented as in STAMP 7. For future use, the interventions can be saved. It prevents the manual input of outliers and breaks altogether.
  - The automatic selection procedure can be repeated with the inclusion of a fixed set of explanatory and intervention variables.

- **Other new features**:

  - Each parameter in the models of STAMP 8 can be edited directly. Parameters can be kept fixed at a particular value. Variances can be kept fixed at values relative to a particular variance of another component (q-ratio). This

facility also applies to multivariate models.
- General forecasting options have been extended and made more flexible. The number of output options for prediction and forecasting have been increased. Future values of explanatory variables available in the database can be used for the forecasting of dependent variables.
- More output diagnostics are presented for predictions (one-step and multi-step), auxiliary residuals and weight and gain functions.

1.5 Developments in STAMP 8.20

- It works under OxMetrics 6.
- Bug fixes.
- The algorithm for the automatic detection of outliers and breaks (option in the Select components dialog) is further advanced and works more effectively.

1.6 The special features of STAMP

(1) *Ease of use*

- STAMP is easy to use, being a **fully interactive and menu-driven** approach to time series modelling: pull-down menus offer available options and dialog boxes provide access to the available functions.
- STAMP has a very **high level of error protection** making it suitable for students or practitioners acquiring experience in time series modelling or with computers, for live teaching in the classroom or fraught late-night research.
- STAMP provides an extensive context-sensitive **help system** explaining both the program usage and the time series methodology.
- The **user friendliness and screen presentations are of the high quality** provided by the OxMetrics system, operating in an edit window to allow documentation of results as analysis proceeds, with easy review of previous results and cutting and pasting within or between windows.

(2) *Advanced graphics*

- STAMP supports **text and graphics together on screen**, with easy adjustment of graph types, layout and colours.
- **Graphs can be documented and edited** via direct screen access with reading from the graph.
- **Time series and cross-plots** are supported with flexible adjustment and scaling options. Spectra, correlograms, histograms and data densities can be graphed.
- **Forecasts and components** can be graphed in many combinations.

1.6 The special features of STAMP

(3) *Flexible data handling*

- The **data handling system provides convenient storage** of large data sets with easy loading to STAMP either as a unit, or for subsamples or subsets of variables.
- **Excel and Lotus spreadsheet files** can be loaded directly, or using 'cut and paste' facilities.
- **Large data sets** can be analysed, with as many variables and observations as memory allows.
- Database variables can be transformed by a **calculator**, or by entering **mathematical formulae** in an editor with easy storage for reuse; the database is easily viewed, incorrect observations are simple to revise, and variables can be documented on-line.
- **Appending** across data sets is simple, and the data used for estimation can be any subset of the data in the database.
- **Several data sets** can be opened simultaneously, with easy switching between the database.

(4) *Model specification and estimation*

- STAMP is designed for **modelling time series** data using models constructed out of interpretable components which are usually specified after the inspection of some of the many graphical displays available in the program. The models are easily setup by clicking onto menus and dialogs in the program.
- The models are **estimated** using maximum likelihood: powerful numerical optimisation algorithms (which exploit analytic derivatives) are embedded in the program with easy user control. Graphical procedures allow the visual inspection of profile likelihood functions.
- Efficient **signal extraction** of time varying components, such as trends and seasonals, is easy. Seasonal adjustment and model-based detrending are available as specific options.
- A **Batch language** allows automatic estimation and evaluation of models, and can be used to prepare a STAMP session for teaching.

(5) *Diagnostic checking and predictive testing*

- **Equation mis-specification tests are automatically provided**. Checks are included on the innovations from the model for residual autocorrelation, heteroskedasticity and normality. More detailed diagnostic checking, including plots of residual density functions is possible. Auxiliary residuals, which are smoothed estimates of the noise in the components of the model, can be used to detect outliers and structural breaks.
- Post-sample and within-sample **predictive testing** is available with full

graphical support, including plots of predictions against outcomes with one-step and multi-step error bars.

(6) *Forecasting*

- It is easy to produce forecasts of the series, together with error bars, and forecasts of trend, seasonal and cyclical **components**.
- Forecasts can be made conditional on future values of the **explanatory variables** which are constructed by assuming a particular growth rate, are fed in manually or are themselves forecasted by building a structural time series model. This makes it easy to forecast under different **scenarios**.

(7) *Output*

- **Graphs can be saved** in several file formats including for later recall, further editing, and printing, or for importing into many popular word processors, as well as directly by 'cut and paste'
- **Results window** information can be saved as an ASCII (human readable) document for input to most word processors, or directly input by 'cut and paste'.
- Model residuals and recursive output can be **stored in the database** for additional graphs or evaluation.

1.7 Basics of the program

1.7.1 Data storage and input

The primary mode of data storage is the IN7/BN7 format. This gives complete compatibility with the other OxMetrics modules. The IN7/BN7 format is based on a pair of files with extensions **.IN7** and **.BN7**. The latter is a binary file containing the actual data, whereas the former holds the information on the contents of the binary file such as variable names and sample periods. The information file is human-readable, whereas the binary file is not. Several real data sets are supplied with STAMP and used in the tutorials.

OxMetrics can read also read other data file formats, including the following spreadsheet files:

- Excel: **.XLS** files;
- Lotus: **.WKS** and **.WK1** files.

Once the data has been entered, they are best stored in IN7/BN7 format. The program checks for potential overwriting of files, and if this is going to occur it offers options for proceeding to overwrite, appending to a file that already exists or selecting another file name. Files may be stored in different directories or drives to those on which the program resides. On input, a search procedure across directories and drives is easily implemented within the Open file dialog if the desired file is not found initially.

The data options allow easy archiving of data. For data samples, reference is by the absolute date in the form *Year Period* to *Year Period,* for example, 1965 1 to 1985 3. Whenever a sample choice has to be made, STAMP will show the maximum available. STAMP can also handle daily data and intra-daily data, see OxMetrics documentation.

1.7.2 Menus and dialogs

The STAMP program is interactive and menu-driven. That is, at each stage a set of options is available, any one of which may be selected. Choices are mainly made by placing the mouse cursor on the menu option and clicking with the left button. It is of course only possible to choose options that make sense: unavailable options are shown in dimmed format. The heart of the STAMP program lies in the **Formulate** and **Test** menus. The dialogs set out the available settings and the user chooses the appropriate configuration. Many dialogs have a default which can be accepted simply by clicking the OK key (using the mouse) or pressing the Enter key ↩. Changing the settings in the dialogs can be carried out using the mouse. In addition, the Tab key can also be used to move the cursor between options. Variables are typically contained in 'List boxes' and the Spacebar plays an important role in that it marks a selection of variables for an operation. The OK button tells the program to carry out the specified operation and/or move on to the next dialog in the sequence, while Cancel returns to the Results or Graphics window.

Other important keys are: the Arrow keys (←↑↓→) and paging keys (PgUp, PgDn, Home and End) which move around windows; the Esc key which cancels instructions or escapes from dialogs; F1 which provides context-sensitive help.

Most operations can be conducted by clicking the left mouse button to select, clicking twice to implement, and clicking and holding down the left button while moving the mouse to drag the cursor (for example, to mark a block of text for cutting and pasting). Any combination of mouse and keyboard is feasible when either would do the job alone.

Several dialogs have list boxes in which a selection can be made, for example to select for a list of variables. Occasionally, only one selection is possible, but often multiple variables can be selected. With the keyboard, you can mark a range of variables by holding the shift key down while using the arrow up or down keys. With the mouse there is more flexibility:

- single click to select one variable;
- hold the left mouse button down to select a range of variables;
- hold the Ctrl key down and click to select additional variables;
- hold the Shift key down and click to extend the selection range.

1.7.3 Help system

STAMP incorporates an extensive, cross-referenced help system which offers advice before crucial decisions are made and can be accessed at any time, either by pressing the F1 key (giving context-dependent help) to call the help menu. If in doubt when using STAMP, press F1 to get help: this facilitates its use as you are much less likely to get stuck. The help system is html-based which means that it works as a webpage. The help system starts the default web-browser on your computer automatically to display the context-dependent help.

Many help, advisory, and warning messages are interspersed throughout the program to pop up as needed. Most of these are self-explanatory queries or comments.

1.7.4 Results storage

All results are shown in the Results window of OxMetrics as calculations proceed, but are not stored on disk unless specifically requested. Results window storage is normally in files with an .OUT extension. Such files can be on a different drive or directory from STAMP. OxMetrics will issue a warning when trying to overwrite an existing file, offering the choice between overwriting the file, appending to the file, or selecting a new destination file name.

1.8 Using STAMP documentation

The documentation comes in four main parts.

I. **Prologue** — introduces the main features provided by the program and sketches how to use it.
II. **Tutorials on Structural Time Series Modelling** — explain how STAMP can be used to model and forecast real data.
III. **STAMP Tutorials** — explain the general working of the program, graphics, data input and modelling.
IV. **Statistical Output** — sets out the statistical details of the models, describes how they are estimated and provides precise definitions of all the statistics and plots in the output.

Assuming the program has been installed, you should proceed as follows after reading the remainder of this chapter:

(a) If you are new to structural time series modelling, follow the example in Chapter 3 to see what the program does at its most basic level.
(b) If you are familiar with the ideas of structural time series modelling, follow the tutorial in Chapter 8.

(c) If you wish to begin by using your own data set, work through the first three tutorials in Part III.
(d) If you would like to read about some of the ways in which STAMP allows you to explore new approaches to modelling economic and financial time series, turn to Chapter 7.
(e) If you want to know about the statistical treatment of the models and the definitions of the diagnostic output, turn to Chapters 9–10.

To use the documentation, either check the index for the subject, topic, menu or dialog that seems relevant, or look up the part relevant to your current activity in the **Contents**, and scan for the most likely keyword. The references point to relevant publications which analyse the methodology and methods embodied in STAMP.

Equations are numbered as (chapter.number); for example, (8.1) refers to equation 8.1, which is the first equation in Chapter 8. References to sections have the form §chapter.section, for example, §8.1 is the first section in Chapter 8. Some sections are starred. These can be left out on first reading for they are concerned with some of the more technical issues.

The typeface conventions adopted in the tutorials are as follows:

(1) Instructions to be typed into the keyboard are shown in `Typewriter` font.
(2) Keyboard commands, including combinations such as `Alt+f`, are also shown in this font.
(3) Menus names are given in **bold,** unless they are just being mentioned in passing.
(4) Dialog names, sometimes appearing as options in menus, are given in Sans serif.
(5) The labels on 'buttons' in dialogs are shown in Sans serif.
(6) The labels used to describe option boxes in dialogs are enclosed in 'raised commas'.
(7) *Italics* are used for technical names.
(8) *Slanted* is used for points to be stressed or comments which are discussing a point not directly related to the main line of argument.

1.9 Citation

To facilitate replication and validation of empirical findings, the STAMP book should be cited in all reports and publications involving its application. The appropriate reference is Koopman, Harvey, Doornik and Shephard (2007).

1.10 World Wide Web

Consult `http://stamp-software.com/` for pointers to additional information relevant to the current and future updates and versions of STAMP.

1.11 Tutorial data sets

STAMP comes with a number of real data sets which are used to illustrate how the program is used for structural time series modelling. All are stored in IN7/BN7 format. This section describes the data sets, in alphabetical order, and discusses some of the questions which they raise. In some cases we indicate how STAMP sets out to answer these questions and take the opportunity to illustrate some of the graphical output of the program.

1.11.1 ENERGY: energy consumption in the UK

This consists of a number of series of quarterly energy consumption (in millions of useful therms). The series are for coal, gas and electricity for the 'Other industries' and 'Other final users' of the economy. The series are given in logged and unlogged form. Thus 'ofuGASl' is the logarithm of gas consumption by Other final users. For further details and a listing of the data, see Harvey (1989).

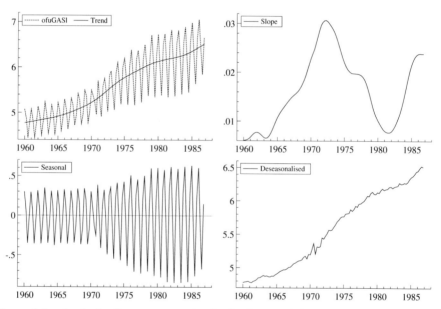

Figure 1.1 Analysis of gas consumption by final users. Original series and estimated: trend, slope of trend, seasonal and seasonally adjusted series.

It is informative to decompose series of this kind into trend, seasonal and irregular components. STAMP does this, and some of its output is shown in Figure 1.1 for 'ofuGASl'. Although the model was formulated in logarithms, the graphs show the implications for the raw data. The first graph shows the trend; there is a dramatic increase in gas usage with the introduction of natural gas from the North sea in the early 1970s. This also shows up in the graph to the right which plots the annualised growth rate of the trend throughout the sample period. The seasonality is shown in the graph in the bottom left hand corner in terms of its multiplicative effect on the trend. Thus at the end of the period it can be seen that, on average, consumption in the winter quarter is running at about 70% above the underlying trend. The greater dispersion in the seasonal pattern over time is due to a higher proportion of gas being used for heating as usage increased in the 1970s. The final graph shows the seasonally adjusted series produced by fitting the structural time series model.

1.11.2 EXCH: daily exchange rates for the US dollar

The four series here are the first differences of the logarithms of daily exchange rates of the dollar against the pound, deutschmark, yen and Swiss franc. The data were recorded at the end of each weekday from 1/10/81 until 28/6/85. Thus the sample size is 946. The interesting issue here is the volatility in the series. Figure 1.2 shows the absolute value of the first difference of the logs of the series together with an estimate of the corresponding underlying volatility in the Deutschmark based on a stochastic volatility model fitted using STAMP.

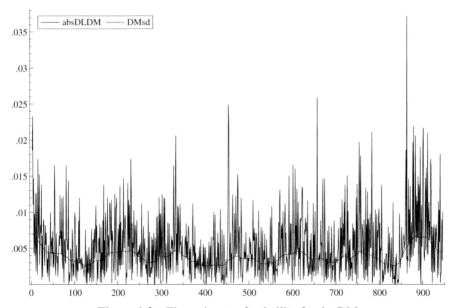

Figure 1.2 The estimate of volatility for the DM.

1.11.3 INTEREST: short- and long-term interest rates

The series 'long' is the monthly yield on 20-year UK gilts. The 'short' series is the 91 day UK Treasury bill rate. Questions to ask are whether the rates contain an underlying component and whether this component emerges if the series are modelled jointly. The data are from Mills (1993, p. 225).

1.11.4 MINKMUSK: minks and muskrats in Canada

The file MINKMUSK contains two series showing the numbers of skins of minks and muskrats traded annually by the Hudson Bay Company in Canada from 1848 to 1909. These data have been studied extensively in the time series literature and as a result provide a useful test-bed for new techniques. There is a known prey-predator relationship between the two species and this gives rise to interlinked cycles. Chan and Wallis (1978) carried out analysis and modelling within a multivariate ARMA framework. They first detrended by quadratic regression on both series. By contrast, STAMP allows the fitting of trend components as part of the overall model. The remaining movements in the series can be captured by a first-order vector autoregression.

1.11.5 NILE: level of the Nile

This is a series of annual observations on the volume of the flow of the Nile in $\times 10^8$ cubic metres; see Balke (1993). Interesting issues which arise include the possibility of a structural break with the construction of the Aswan dam in 1899 and the existence of outliers. A cycle might also be present.

1.11.6 RAINBRAZ: rainfall in north-east Brazil

This is a series of annual observations on rainfall in Fortaleza in North-East Brazil, starting in 1849. The unit of measurement is centimetres, with the data being recorded to the nearest millimetre. The series is one of the longest records of tropical rainfall available, and it has been subject to a great deal of study since the region inland from Fortaleza frequently suffers from severe droughts. The main issue concerns the existence of a cycle, or cycles, in the pattern of rainfall; see, for example, Kane and Trivedi (1986).

1.11.7 SEATBELT and SEATBQ: road casualties in Great Britain and the 1983 seat belt law

The SEATBELT data consists of monthly observations on the numbers of drivers, front seat passengers and rear seat passengers who were killed or seriously injured (KSI) in road accidents in cars in Great Britain. Data on the number of kilometres travelled and the real price of petrol is also included. Harvey and Durbin (1986) used these data to carry out an extensive study of the effect of the seat belt law of January 31st, 1983.

The file SEATBQ contains only the last observation in each quarter, (that is March, June, September and October) in order to make the analysis more manageable.

The data provide a good illustration of the way in which the intervention analysis of the effect of the seat belt law can be extended by using control groups.

1.11.8 SPIRIT: consumption of spirits in the UK

The SPIRIT file contains the per capita consumption of spirits in the UK from 1870 to 1938, together with per capita income and relative price; see Prest (1949). The consumption of spirits can only be partly explained by income and price, presumably because of a trend component representing changes in tastes and social conditions. Such variables cannot be measured, but including a deterministic time trend leaves considerable serial correlation in the residuals. The solution is to include a stochastic trend component in the regression model.

1.11.9 TELEPHON: telephone calls to three different countries

This file contains data on the logarithms of paid minutes of telephone calls from Australia to three different countries.

1.11.10 UKCYP: consumption, income and prices in the UK

The data on quarterly UK real personal disposable income (Y), consumer non-durables (C) and the GDP deflator (P) can be found in the file named UKCYP. Lower case letters denote the variables in logarithms.

1.11.11 USYCIMP: US macroeconomic time series

This file consists of data on quarterly, seasonally adjusted, logarithms of observations on three real, per capita US series. These are 'private' GNP (that is, not including government expenditure), consumption and investment, denoted 'y', 'c' and 'i' respectively. In addition, the file includes the price level as represented by the GNP deflator, 'p', and 'm', the money supply, or, more precisely, M2; see King, Plosser, Stock and Watson (1991).

The first question which often arises with such economic data is what are its main features, particularly with respect to cyclical behaviour. Economists sometimes refer to such features as 'stylised facts'. STAMP offers one possible way of obtaining a set of stylised facts. For example, we can fit a model consisting of a trend and a cycle to GNP. Figure 1.3 shows the result of fitting such a model and its implications for forecasting. The data are in logarithms and the forecasts are for five years from the end of 1985. The final panel shows the forecasts with a band of one RMSE on either side.

18   *Chapter 1 Introduction*

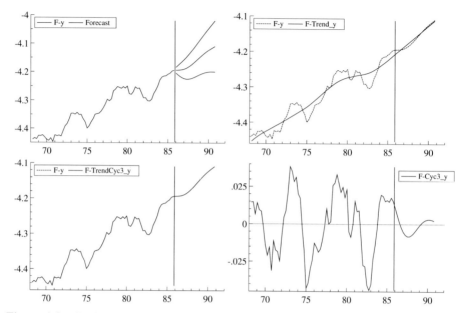

Figure 1.3 Stylised facts of US GNP. Series and estimated trend and cycle. Forecast and forecast confidence intervals.

1.11.12 USmacro07: more recent US macroeconomic time series

This file contains the latest US data on key variables and essentially updates the previous file. It contains quarterly data on real U.S. Gross Domestic Product, Investment and Consumption from 1947(1) to 2007(2), obtained from the Department of Commerce (website: www.bea.gov). The logarithms are LGDP, LINV and LCONS. The Consumer Price Index CPI is from the U.S. Bureau of Labor Statistics (website: www.bls.gov) and the series INFLcpi is the (annualized) rate of inflation, measured as the first differences of the quarterly CPI multiplied by four hundred. INFLdef is constructed in the same way but from the GDP deflator. Unlike the real series, CPI is not seasonally adjusted.

1.12 Data sets used in exercises

The following data sets are used in the exercises which appear at the end of some of the tutorials:

AIRLINE: This is a well-known data set consisting of the number of UK *airline passengers* in thousands from January 1949 to December, 1960; see Box and Jenkins (1970).

EMPL: The two series in this file, 'empl' and 'output', are the logarithms of *employment and output in UK manufacturing*. The data are seasonally adjusted and employment is measured in thousands while output is an index with the value in 1983 set to 100. The dynamic employment-output equation needs a stochastic trend component to allow for changes in productivity; see Harvey, Henry, Peters and Wren-Lewis (1986).

ICEVOL: This data set consists of 220 points, corresponding to intervals of 2,000 years, on oxygen-18 levels measured from deep sea cores. These measurements act as proxies for *global ice volume*; see Newton, North and Crowley (1991). One reason for being interested in forecasting from this data set is the question of whether greenhouse effect warming could be offset by a future ice age. The data are thought to contain at least three cycles, but it is an open question as to whether or not they are strictly deterministic.

LAXQ: This consists of quarterly observations of *US exports to Latin America*; see Bruce and Martin (1989). There is the possibility of outliers due to dock strikes.

PURSE: This file contains the number of *purses snatched* in the Hyde Park area of Chicago each lunar month. Since the data consist of small integers, it is interesting to explore transformation, such as taking square roots, before a model is fitted.

1.13 STAMP and PcGive

STAMP shares the same structure as PcGive with regard to data handling and the output of results. Furthermore the conventions adopted in the modelling dialogs are similar. As a result users of PcGive are not only able to use data files directly with STAMP, but will also find the program easy to use.

Despite the superficial similarities, the analysis and modelling carried out by STAMP is quite different to that in PcGive. The programs are designed for different purposes and are essentially *complementary*. PcGive is primarily an econometrics packages, while STAMP addresses a wide range of time series applications in many disciplines. Indeed the structural time series approach which it is designed to implement is much more of a competitor to the ARIMA approach originally popularised by Box and Jenkins (1970). The case for using structural time series models rather than ARIMA models is set out at some length in Harvey (1989), West and Harrison (1997) and Durbin and Koopman (2001).

PcGive is primarily a program for carrying out single and multiple equations regression, but with the option of using techniques such as instrumental variables and non-linear least squares. STAMP will also carry out regression, but this is not its prime purpose and for serious economic work, PcGive is to be preferred for its more comprehensive facilities and output. PcGive enables the user to carry out some univariate time series analysis, but this is within an autoregressive framework, stressing the role of unit root tests. It is not the same as STAMP's modelling and analysis of components.

Chapter 2
Getting Started

We now discuss some of the basic skills required to get started with STAMP. We will give a short introduction to the services provided by OxMetrics. OxMetrics provides the data which STAMP analyses, and receives all the text output and graphs which you create in STAMP. When you start STAMP, it automatically starts up OxMetrics. Note that STAMP can also be started directly from OxMetrics. The aim of this tutorial is just to load data and create graphs: for instructions on how to transform data using the calculator or algebra, consult the OxMetrics book.

2.1 Starting STAMP

First start OxMetrics, then STAMP can be started from the Modules menu of OxMetrics:

Alternatively, start STAMP from the taskbar or from the OxMetrics group. If this is the first time you have used STAMP, your initial screen could look like the capture shown below.

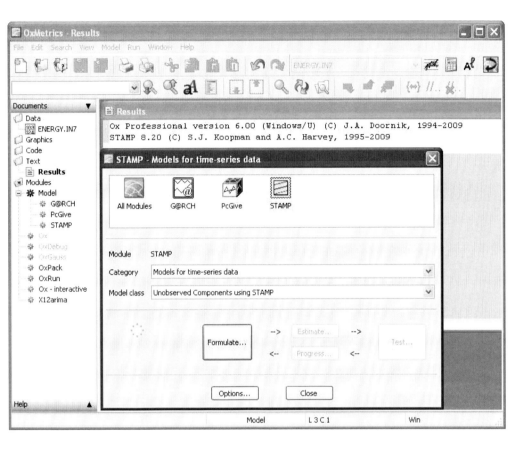

2.2 Loading and viewing the tutorial data set

Without data, STAMP cannot operate, so the first step is to load data into OxMetrics which STAMP can then access. Here we load the energy consumption data set, which was introduced in §1.11.1. The data set is called ENERGY.IN7, and it contains a number of series on quarterly energy consumption in the UK. The IN7 file is a human-readable file, describing the data. There is a companion BN7 file, which holds the actual numbers (in binary format, so this file cannot be edited). OxMetrics can handle a wide range of data files, among them Excel (.XLS) and Lotus files (.WKS and .WK1), and of course plain human-readable (ASCII) files. You can also cut and paste data between Excel and OxMetrics. Details are in the OxMetrics book.

To load the tutorial data set, access the File menu in OxMetrics:

22   *Chapter 2 Getting Started*

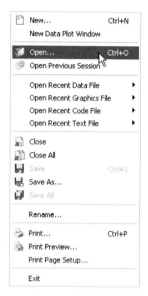

and choose Open. If you installed in the default directory structure the data will be in the \Program Files\STAMP directory, so locate that directory and select ENERGY:[1]

---

[1] You may or may not see the .IN7 file extension of the data files. This depends on the settings in the Explorer options. ENERGY does have a little OxMetrics picture, to indicate that the file type is associated with OxMetrics.

The data file will be loaded, and displayed in OxMetrics:

ENERGY.IN7 - D:\OxMetrics4\data\ENERGY.IN7					
	ofuEL	ofuEL1	ofuGAS	ofuGAS1	oiGAS
1960(1)	188	5.23644	160.1	5.0758	190.8
1960(2)	149.8	5.0093	129.7	4.86522	174.6
1960(3)	103	4.63473	84.8	4.4403	146.4
1960(4)	157.2	5.05752	120.1	4.78832	177.5
1961(1)	202.9	5.31271	160.1	5.0758	187.9
1961(2)	159.3	5.07079	124.9	4.82751	171.6
1961(3)	123.2	4.81381	84.8	4.4403	143.5
1961(4)	166.8	5.1168	116.9	4.76132	174.6
1962(1)	231.6	5.44501	169.7	5.13403	193.8
1962(2)	190.1	5.24755	140.9	4.94805	173.1
1962(3)	138.1	4.92798	89.7	4.49647	143.5
1962(4)	185.9	5.22521	123.3	4.81462	174.6
1963(1)	268.7	5.5936	187.3	5.23271	199.7
1963(2)	208.2	5.3385	144.1	4.97051	171.6
1963(3)	148.7	5.00193	92.9	4.53152	145

Double clicking on the variable name shows the documentation of the variable if any is available. It would also allow renaming the variable. The data can be manipulated, much like in a spreadsheet program. Here we shall not need these facilities. Do not click on the cross of this window: this closes the database, thus removing it from OxMetrics and so from STAMP.

2.3 OxMetrics graphics

The graphics facilities of OxMetrics are powerful yet easy to use. This section will show you how to make time-plots and cross-plots of variables in the database. OxMetrics offers automatic selections of scaling etc., but you will be able to edit these graphs, and change the default layout such as line colours and line types. Graphs can also be saved in a variety of formats for later use in a word processor or for reloading to OxMetrics.

2.3.1 A first graph

Graphics is the first entry on the Model menu. It can also be activated by clicking on the cross-plot graphics icon on the toolbar:

Activate the command to see the following dialog box:

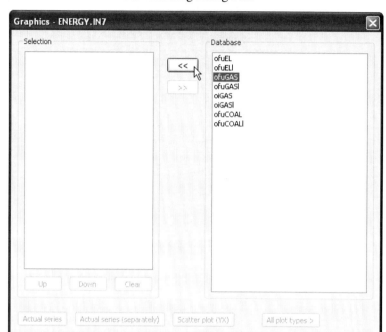

This is the first example of a dialog with a multiple selection list box. In such a list box you can mark as many items as you want. Here we select the variables we wish to graph. With the keyboard you can only mark a single variable (by using the arrow up and down keys) or range of variables (hold the shift key down while using the arrow up or down keys).

With the mouse there is more flexibility:

- single click to select one variable;
- hold the left mouse button down to select a range of variables;
- hold the Ctrl key down and click to select additional variables;
- hold the Shift key down and click to extend the selection range;

In this example we select *ofuGAS*, as shown in the capture above, and then press the << button. Next select *ofuGASl*, press << and, finally, the button Actual series (separately). The graph which appears looks very much like Figure 2.1. It confirms the claim made in §1.11.1, that the seasonality is much more constant after taking logarithms.[2] Note that there are many on-screen edit facilities. For example, you can move the position of the legend by picking it up with the mouse. The OxMetrics book describes the edit facilities in more detail.

---

[2] With logarithm we always mean the natural logarithm, denoted $\log(\cdot)$, unless a different base is explicitly given.

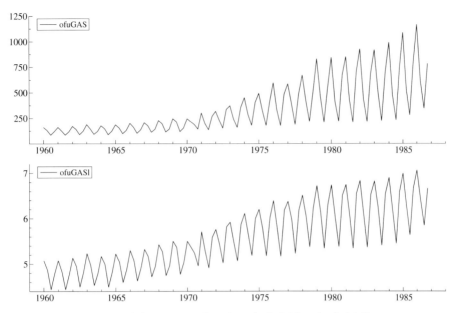

Figure 2.1 Time-series plot of *ofuGAS* and *ofuGASl*.

2.3.2 Graph saving and printing

To print a graph directly to the printer, click on the printer icon in OxMetrics. You can preview the result first using the Print Preview command from the File menu.

Graphs can be saved to disk in various formats:

- Enhanced metafile (.EMF);
- The MS Word print format (.PNG);
- Encapsulated PostScript (.EPS), which is the format used to produce all the graphs in this book;
- PostScript (.PS), this is like EPS, but defaulting to a full page print.
- OxMetrics Graphics File (.GWG).

When you save a graph in any format, the GWG file is automatically saved alongside it. Then, when loading a previously saved EPS file (say), OxMetrics can use the GWG file to reload the actual graph.

2.4 Data transformations

To complete this short introduction to OxMetrics, we do some data transformations using the OxMetrics Calculator. Section 1.11.2 introduced the EXCH data set. Load this data set into OxMetrics. The status line at the bottom of OxMetrics now indicates that this is the active database (to change between databases, change the selection in the toolbar).

The objective is to create the logs of the first differences of the $/£ exchange rate. Activate the Calculator, select Pounds from the list of variables, and click on the log button. The expression log(Pound) will be written in the expression window:

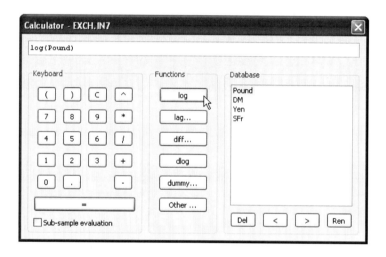

Press Enter (or click on the = button). You will be prompted for a name:

Accept the default. Now use the diff button to take first differences of *LPound* in a similar way. This creates *DLPound*. Finally, to take the absolute values, select *DLPound*, then press the button Other... locate the function in the Functions dailog, and double click on fabs:

2.4 Data transformations

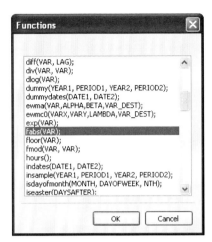

The expression reads `fabs(DLPound)`. Press =, and give the variable a name, *absDLPound* for example. The transformations were logged in the Results window:

```
Algebra code for EXCH.IN7:
LPound = log(Pound);
DLPound = diff(LPound,1);
absDLPound = fabs(DLPound);
```

This code can be saved, and rerun later. More information on Algebra is in the OxMetrics book. Figure 2.2 shows the created absolute values of the $/£ exchange rates in the upper part, together with a spline line (obtained using automatic bandwith selection, see the Cross Plot page in the OxMetrics Graphics dialog). Figure 2.2 plots in the bottom part the sample autocorrelation function of the absolute values.

This completes the getting started chapter. We hope that you're equipped now for more substantial structural time series modelling exercises.

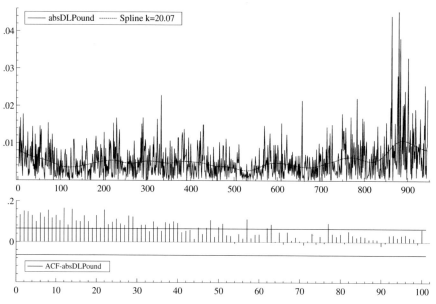

Figure 2.2 Time series plot of *absDLPound* and sample ACF of *absDLPound*.

Part II

Tutorials on Structural Time Series Modelling

Chapter 3
Introduction to Univariate Modelling

This chapter provides a simple introduction to structural time series modelling using STAMP. A recent introductory textbook treatment for this class of time series models is provided by Commandeur and Koopman (2007). Most attention in this chapter is focused on the **Formulate** and **Test** menus of STAMP. The previous chapter showed how to start STAMP, and load the ENERGY data set in OxMetrics.

Prior to modelling we graph the 'ofuCOAL' series in OxMetrics, see Figure 3.1. The graph of this series, which is quarterly consumption of coal by 'Other final users' shows a clear downward trend and a seasonal pattern which seems to decrease over time. The downward trend is still apparent when the logarithm of the series, 'ofuCOALl', is graphed. However, the seasonal pattern appears more stable and so, as in many cases, the log transform is to be preferred.

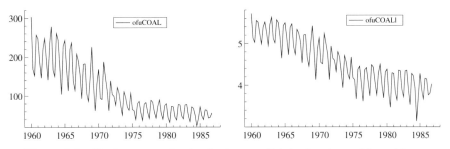

Figure 3.1 Coal consumption by final users. Original series and logarithms.

If you wish, you can explore more transformations and examine the correlograms of the observations after various types of differencing. This can be done using the tools provided by Calculator and Algebra and using OxMetrics Graphics in the Model menu.

3.1 Model formulation

Structural time series modelling in STAMP starts with the Model dialog. Here, the variable(s) to be modelled are selected, and additional model components specified.

After successful estimation, model evaluation proceeds from the **Test** dialog.

The Model dialog can be activated in several ways. For example, it can be selected from the **Model** menu in OxMetrics.

You can also click on the Model icon in the OxMetrics toolbar:

Finally you can select it from Modules in the Documents window or just by pressing Alt+y:

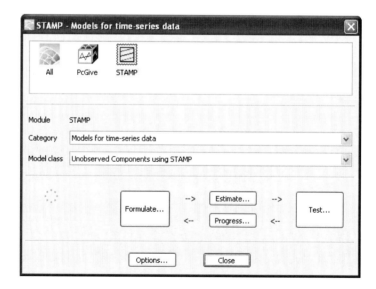

Inside the Model dialog, choose Formulate by clicking on its button. The Formulate dialog allows you to mark 'ofuCOAL1' by using the mouse and press the << button (or double-click on the variable). You will see:

3.1 Model formulation

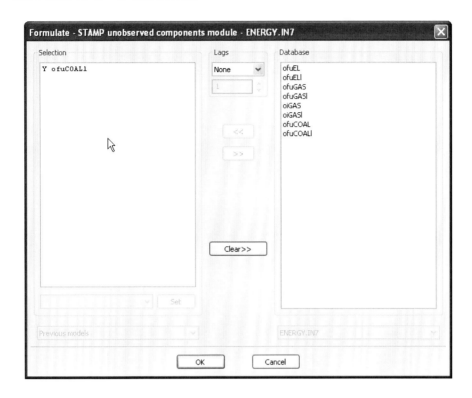

The 'ofuCOALl' series has now appeared under 'Selection' as a Y, or dependent, variable. As we will be using no explanatory variables in this chapter, we are ready to proceed. Press OK, or Enter since OK is highlighted.

The next stage is to select a suitable set of components, on the basis of your knowledge of the salient features of the series. Most quarterly economic series display trend and seasonal movements and in the case of ofuCOALl such movements were apparent from the graph.

The Select components dialog lists the available options:

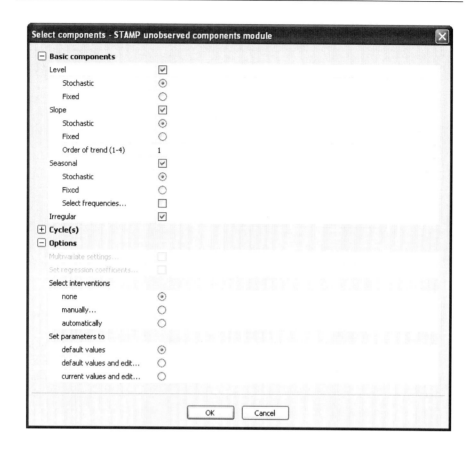

The default setting in the Select components dialog is the *Basic Structural Model* (BSM). This specifies a 'Stochastic Level' and 'Stochastic Slope', making up the trend, a 'Stochastic Seasonal' and an 'Irregular'. These settings may be changed by moving around the dialog using the Tab or Arrow keys (up and down only in this case) and marking the desired box with the Spacebar. Alternatively, click on the desired box with the mouse. For the moment accept the BSM by choosing OK; you can simply hit ↵ since the OK button is highlighted; this leads to the Estimate Model dialog. It allows you to change the sample period if you wish. The other options in the dialog are for different estimation methods and may be ignored at this stage. Press Enter or click OK.

Once estimation is complete, some basic information appears in the Results window. In order to read everything you may need to move up or down the screen using the Arrow keys or the mouse. Using the default options, the output will read:

3.1 Model formulation

```
UC( 1) Estimation done by Maximum Likelihood (exact score)
    The selection sample is: 1960(1) - 1986(4) (T = 108, N = 1)
    The dependent variable Y is: ofuCOAL1
    The model is:  Y = Trend + Seasonal + Irregular
    Steady state.......... found without full convergence

Log-Likelihood is 188.015 (-2 LogL = -376.031).
Prediction error variance is 0.0205058
Summary statistics
                  ofuCOAL1
T                 108.00
p                   3.0000
std.error           0.14320
Normality           6.6991
H(34)               1.5112
DW                  1.8476
r(1)                0.063574
q                  12.000
r(q)               -0.063351
Q(q,q-p)            6.0221
Rs^2                0.35124
```

```
Variances of disturbances:
                    Value        (q-ratio)
Level           0.000749318    (  0.04554)
Slope           2.75126e-006   (0.0001672)
Seasonal        0.000000       (  0.0000)
Irregular       0.0164540      (  1.000)

State vector analysis at period 1986(4)
                         Value        Prob
Level                  3.90276    [0.00000]
Slope                 -0.00685    [0.36652]
Seasonal chi2 test   485.58515    [0.00000]
Seasonal effects:
              Period     Value        Prob
                  1    0.27043    [0.00000]
                  2   -0.09705    [0.00002]
                  3   -0.40679    [0.00000]
                  4    0.23342    [0.00000]
```

The 'Estimation report' tells us that convergence was Very Strong. This is good news. Maximum likelihood estimation has just been carried out by numerical optimisation and STAMP is telling us that this was successful. A failure to satisfy the various convergence criteria may be an indication of a poorly specified model

The 'Diagnostic summary report' provides some basic diagnostics and goodness-of-fit statistics. In particular, we have the Box–Ljung Q-statistic, $Q(12,9)$. This is a test for residual serial correlation, which is based on the first 12 residual autocorrelations and should be tested against a chi-square distribution with 9 degrees of freedom. It is not significant here. If you wish to find the 'p-value', go to the OxMetrics Model/Tail Probability menu. Select 'Chi^2(n1)' in the dialog and enter '9' for 'n1' and the value of the test statistic in 'value'. The result is Chi^2(9) = 6.0221 [0.7377].

Since the estimation procedure converged and the diagnostics appear satisfactory, we can be reasonably confident that we have estimated a sensible model (though it may not be the best).

The variances govern the movements in the components. They are part of the standard model output, and were already printed in the Results window:

```
Variances of disturbances:
                    Value        (q-ratio)
Level           0.000749318    (  0.04554)
Slope           2.75126e-006   (0.0001672)
Seasonal        0.000000       (  0.0000)
Irregular       0.0164540      (  1.000)
```

Interpreting the actual numbers is not easy for the inexperienced user and, in fact, the information they contain is effectively displayed in the plots of the estimated components. However, a *zero* parameter estimate does convey information since it tells us that the corresponding component is fixed. Thus in the present example the seasonal pattern is fixed. As part of the default output, the components graphics can be viewed from the Documents window, in Graphics/Model, see Figure 3.2.

3.1 Model formulation

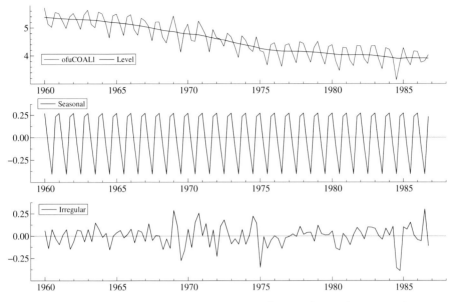

Figure 3.2 Components graphics output of model for coal consumption.

To investigate the parameter estimates more extensively, more detailed diagnostics and forecasts can be accessed from the **Model** dialog by clicking on the **Test** button. The **Test** dialog appears.

3.2 Evaluating and testing the model

The **Test** dialog contains eight options. The first three are concerned with estimates of parameters and components, the fourth and sixth give additional diagnostics, while the fifth allows the user to carry out diagnostic checking by auxiliary residuals. The penultimate option is for forecasting, and is described in the next section. The last option is for the storage of output in the OxMetrics database. The listed order is quite logical, though the easiest way to determine how well the model is working is often to move straight to Components graphics and examine the various plots.

To access the different **Test** dialogs, one or more options can be selected and click on OK.

3.2.1 More written output

The final state vector contains information on the values taken by the various components at the end of the sample. To see the final state values, access More written output on the **Test** menu, and make sure that the options 'State vector analysis' and 'State and regression output' are checked:

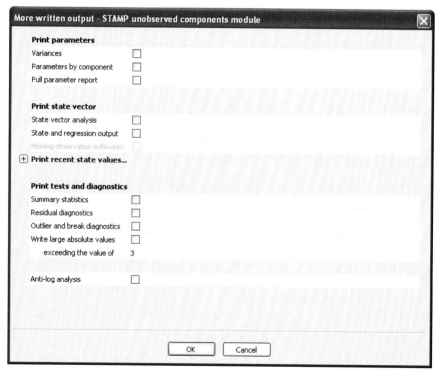

This prints results for the final state vector:

3.2 Evaluating and testing the model

```
State vector analysis at period 1986(4)
                          Value       Prob
Level                   3.90276   [0.00000]
Slope                  -0.00685   [0.36652]
Seasonal chi2 test    485.58515   [0.00000]
Seasonal effects:
             Period     Value       Prob
                  1    0.27043   [0.00000]
                  2   -0.09705   [0.00002]
                  3   -0.40679   [0.00000]
                  4    0.23342   [0.00000]
Equation ofuCOAL1: coefficients of components in final state at period 1986(4)
             Coefficient      RMSE     t-value        Prob
Level            3.90276   0.06218   62.76158   [0.00000]
Slope           -0.00685   0.00755   -0.90701   [0.36652]
Seasonal         0.16524   0.01768    9.34847   [0.00000]
Seasonal 2       0.33861   0.01768   19.15706   [0.00000]
Seasonal 3       0.06818   0.01242    5.48945   [0.00000]
```

Thus, for example, the slope is -0.00685 and the figure in square brackets after the 't-value' is a two-sided Prob. value; that is, it shows the probability of getting an absolute value of a standard normal variable greater than this value if the true parameter is zero.

In our particular example, as in many other cases, the data are in logarithms. Returning to the More written output menu, we now check the 'Anti-log analysis' option, together with 'State vector analysis':

```
State vector anti-log analysis at period 1986(4)
It is assumed that time series is in logs.
                             Value       Prob
Level (anti-log)          49.53886   [0.00000]
Level (bias corrected)    49.63473   [      -]
Slope (yearly %growth)    -2.73858   [0.36652]
Seasonal chi2 test       485.58515   [0.00000]
Seasonal effects:
             Period         Value       Prob       %Effect
                  1       1.31052   [0.00000]     31.05223
                  2       0.90751   [0.00002]     -9.24935
                  3       0.66578   [0.00000]    -33.42185
                  4       1.26291   [0.00000]     26.29149
```

The additional information shows, for example, that the slope can be interpreted as a growth rate and we see that here the estimated current growth rate is -2.74% per year. The seasonals can be interpreted as the factors by which we multiply the trend, so that, for example, consumption of coal is, on average, 31% higher in the winter, Seas 1.

3.2.2 Components graphics

The information provided in Components graphics is fundamental to the interpretation of the model. The default, shown on the next page, provides what we consider to be the most useful plots, but others may be obtained by marking the required boxes in the usual way.

The *smoothed* estimates of the components are obtained using all the information in the sample; that is, they are constructed using observations which come after as well as those which come before. This is sometimes referred to as *signal extraction*.

When the data are in logarithms it is often helpful to look at the trend and seasonal components after exponentiating (anti-logs). To do this mark 'Anti-log analysis'. The interpretation of the seasonals and growth rate graphs, see Figure 3.3 is as in 'State vector analysis' except that now we see any changes which may have taken place over the whole sample period. However, in the present example both are fixed.

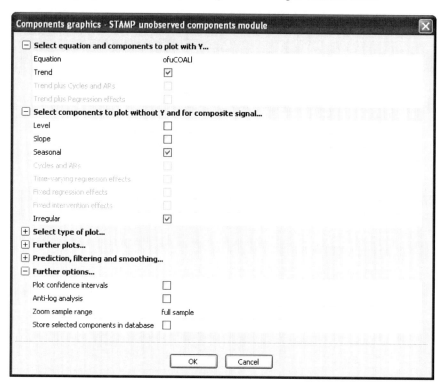

3.2.3 Weight functions

Structural time series models belong to the class of linear time series models. Therefore, the smoothed estimates of the components are linear functions of all observations in the selected sample. It may be of interest to investigate the actual obervation weights that are used for the computation of the smoothed estimate at a particular time point within the sample. To facilitate such an analysis, the dialog Weight functions allows, along more specialised output, the graphical representation of the weights. To illustrate this, access the dialog Weight functions:

3.2 Evaluating and testing the model

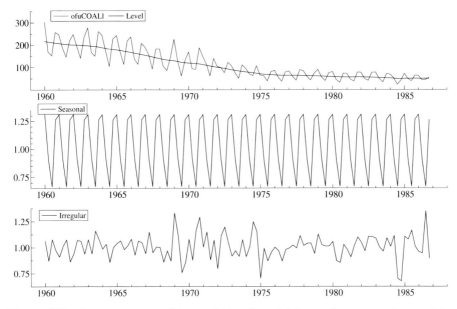

Figure 3.3 Components graphics analysis of model for coal consumption, anti-log analysis.

The default setting produces the graphical output as presented in Figure 3.4 for the estimated level component at period 1973 Q2. It is clear that the weights decay exponentially with the distance of the observations from the period 1973 Q2. Further the symmetric distribution of the weights around period 1973 Q2 is a feature that is common to linear and time-invariant structural time series models.

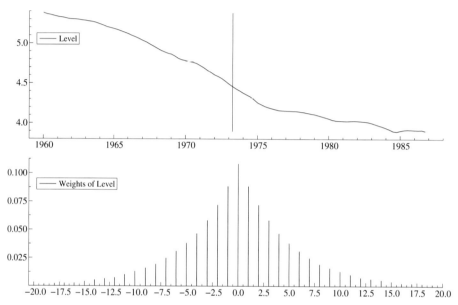

Figure 3.4 Weight graphics analysis for trend of model for coal consumption at period 1973 Q2.

3.2.4 Residuals graphics

Further diagnostic checking of the model may be carried out by plotting the residuals and looking at their full correlogram. A graph of the distribution of the residuals can also be constructed. These options are provided through the Residuals graphics dialog.

The default produces Figure 3.5. Additional output can be requested through the option 'Write diagnostic statistics' and it outputs an extensive list of residual, goodness-of-fit and serial correlation statistics:

3.2 Evaluating and testing the model

```
Normality test for Residuals ofuCOAL1
Sample size         103.00
Mean              0.024851
St.Dev            0.99969
Skewness         -0.59764
Excess kurtosis   0.99695
Minimum          -3.0098
Maximum           2.4635
                  Chi^2       Prob
Skewness          6.1315    [ 0.0133]
Kurtosis          4.2656    [ 0.0389]
Bowman-Shenton   10.397     [ 0.0055]

Goodness-of-fit based on Residuals ofuCOAL1
                                            Value
Prediction error variance (p.e.v)          0.020506
Prediction error mean deviation (m.d)      0.015444
Ratio p.e.v. / m.d in squares              1.1223
Coefficient of determination R^2           0.94281
... based on differences Rd^2              0.89242
... based on diff around seas mean Rs^2    0.31974
Information criterion Akaike (AIC)        -3.7759
... Bayesian Schwartz (BIC)               -3.6269
Serial correlation statistics for Residuals ofuCOAL1
```

```
Durbin-Watson test is 1.86205
Asymptotic deviation for correlation is 0.096225
    Lag      df      Ser.Corr    BoxLjung       prob
     4        1     -0.0089215     1.4639     [ 0.2263]
     8        5     -0.0052752     2.6772     [ 0.7496]
    12        9     -0.063513      6.3039     [ 0.7092]
```

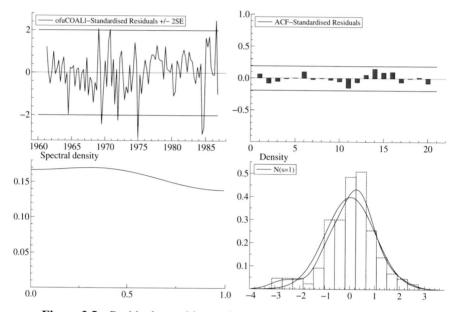

Figure 3.5 Residuals graphics analysis of model for coal consumption.

The Auxiliary residuals option is concerned with the detection of outliers and structural change. It will be described when interventions are dealt with in Chapter 5.

3.2.5 Prediction graphics

Prediction graphics is for making predictions as if data has been held back at the end of the sample. The number of observations to withhold can be specified in the dialog. In the illustration below, we change the default from 8 to 12. The Prediction graphics dialog can be activated via the **Test** menu:

3.2 Evaluating and testing the model 45

Prediction graphics - STAMP unobserved components module

Select equation and type of predictions
Equation   ofuCOAL1
One-step ahead   ◉
Multi-step ahead   ○
Post-sample size   12

Plot predictions and Y
Predictions   ☑
... with Y   ☑
... with standard errors   ☑
... and scaled by   2
Cross-plot predictions x Y   ☑

Plot residuals
Residuals   ☐
... with standard errors   ☐
... and scaled by   2
Standardized residuals   ☑
Cumulative sum   ☑
Cumulative sum t-test   ☐

Write
prediction tests   ☐

[OK]   [Cancel]

Apart from changing the post-sample size to 12, accepting the default produces graphs as shown in Figure 3.6. In the first graph it can be seen that the values of 1984 Q3 and Q4 are outside the prediction intervals, set at two root mean square errors (RMSEs). The three subsequent predictions are also poor. The forecast Chow test can also be given (in addition to the 12 post-sample predictions):

```
Equation ofuCOAL1: post-sample predictive tests.
Failure Chi2( 12) test is   29.4566  [0.0034]
Cusum t( 12)      test is    0.2668  [0.7941]
```

The figure shown in square brackets indicates that the probability of a value greater than this magnitude is 0.0034. This provides further evidence that the prediction errors are not consistent with the model.

The above exercise may be repeated with 'Multi-step prediction' marked and with post-sample size set to 12. In this case the predictions are made using the information at the end of 1983 and are not updated with the arrival of each new observation. The result is that although Q3 and Q4 of 1984 are again poorly predicted, the next three observations are predicted very well. We therefore conclude that Q3 and Q4 of 1984 are unusual observations, and in fact this turns out to be the case since the coal miners were on strike then. Hence the model is a good one and what needs to be done is to remove the two strike observations; see Chapter 5 on intervention analysis.

46   *Chapter 3 Introduction to Univariate Modelling*

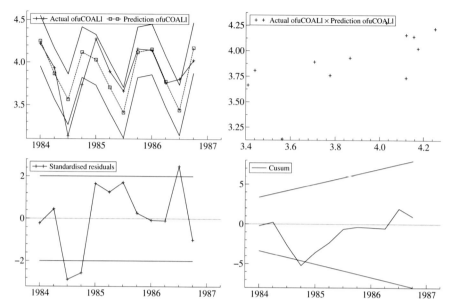

Figure 3.6 Prediction analysis of model for coal consumption.

3.2.6 Forecasting

The Forecasting dialog is for making predictions over a pre- or post-sample period; that is, for a period where no observations on the series are available. The dialog allows us to look at forecasts of components as well as forecasts of the series itself. Just mark the appropriate boxes.

The lines in Figure 3.7 indicate *one* RMSE on either side. Note that the prediction interval increases the further forward we move into the future.

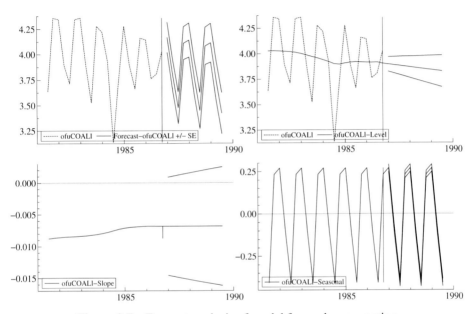

Figure 3.7 Forecast analysis of model for coal consumption.

3.3 Exercises

(1) The remaining series in the ENERGY file illustrate other interesting features. In interpreting the GAS series, it needs to be borne in mind that in 1970 cheaper natural gas became available from the North Sea. The jump in consumption is quite dramatic in 'Other industries' — oiGAS — but STAMP copes with it very well. Show this by fitting a BSM to the logged series. Make forecasts up to five years ahead.

(2) Return to ofuCOAL, but now mark 'Fixed level' in the Components dialog. Show that the result is a much smoother trend and a growth rate which gradually changes over time. Compare the overall fit to that of the unrestricted model. Do you think that this 'smooth trend' specification has more attractive properties?

(3) Fit a model to the electricity series, ofuELECl. Comment on the movements in the trend and seasonal. Compare with the results from ofuGASl in §1.11.1. Carry out post-sample predictive testing over the last two years. What is the value of the Chow statistic?

Chapter 4

Tutorial on Components

4.1 Selection of components

The specification of components is at the heart of structural time series modelling. This section sets out the methodology underlying the meaning and selection of components. The next four sections give details of the various components and provide examples showing how a particular model is specified and evaluated. The statistical specifications of the various components are in separate subsections and a degree of understanding of what the models can achieve is possible without fully understanding them.

Initial specification — An initial judgement about components can often be made on the basis of prior knowledge of the series. For example, a seasonal component will typically be included if the observations are quarterly. A graph of the data often provides confirmation or further information. The various options in OxMetrics allow the exploration of various transformations, such as logarithms. The correlograms of series which have been made stationary by differencing can also be examined, although this type of analysis plays a much less prominent role than it does in the ARIMA 'Box-Jenkins' model selection methodology.

Estimation — Once a model has been specified, it is estimated. If convergence problems arise, this may be an indication of a poor specification or too many parameters.

Parameters — The variance ratios govern the extent to which the components move stochastically. The actual numbers are not easy to interpret. However, a value of zero for a variance indicates that the corresponding component is deterministic. If this is the case, a standard regression type significance test can be carried out on the corresponding component in the state. If it is not significantly different from zero, it may be possible to simplify the model by eliminating that particular component.

Components — In general, the components are stochastic and so can only be assessed by looking at their behaviour throughout the whole sample, rather than at the end. The behaviour of the smoothed components provides a guide as to whether the decomposition implied by the fitted model is useful.

Diagnostics — Once the model has been estimated, various diagnostic and goodness-of-fit statistics are printed out. Further diagnostic checking of the model as a whole can be carried out by accessing **Test**/Residuals graphics dialog.

Predictive testing — Predictions near the end of the series give a good idea of the properties of model and how well it fits. STAMP offers the option of computing and displaying predictions and carrying out predictive tests when data is held back at the end of the sample. These tests are based on one-step ahead predictions. However, it is often informative to examine the behaviour of extrapolative (multi-step) predictions as well.

If we try out many different specifications, there is the danger of data mining. By retaining some observations at the end of the series for post-sample predictive testing, we guard against a spuriously good fit. These post-sample observations are not used to select the model or to estimate the parameters.

4.2 Trend

The trend is the long-run component in the series. It indicates the general direction in which the series is moving.

There are two parts to the trend:

- *Level* — the actual value of the trend;
- *Slope* — this component of the trend may or may not be present.

In order to illustrate the various aspects of trend specification, we will use the data set of US macroeconomic time series — USYCIMP.IN7. These logged observations are all quarterly, but seasonally adjusted. If you are not in STAMP, you may start the program within OxMetrics from the menu **Model** or by pressing `Alt+y`. The database can be loaded into OxMetrics by using the option **File**/Open.

4.2.1 Local level model

The local level model consists of a random disturbance term around an underlying level which moves up and down, but without any particular direction.

We start with the rate of inflation, which is denoted as Dp. It is the first difference of the logarithm of the price level, p. From the **Model**/Graphics dialog in OxMetrics, you can produce a graph of the series, see Figure 4.1. As you will see, it has the characteristics of a local level model.

To estimate a local level model for Dp, go to the **Model**/Formulate dialog of STAMP and select Dp from the Database listbox bu using your mouse and press $<<$ (or double-click on Dp). If you are continuing from the previous section, change the database by selecting the USYCIMP database below the Database listbox (of course, this database must be loaded into OxMetrics first). If Dp appears as an Y variable in the Selection listbox, press OK to go to the Select components dialog. Here we remove the slope component from the model by clicking on the checkbox of 'Slope' (the cross sign

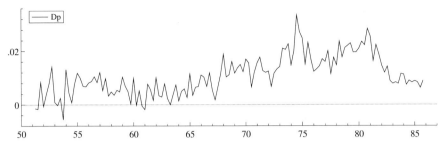

Figure 4.1 Time series of US inflation, from data set USYCIMP.

disappears) and we repeat this for the seasonal component. We keep the other default options. We have specified the local level model. By clicking on OK we enter the Estimate dialog in which we can keep all the defaults and only require to click on OK. The model is estimated and output is reported in the Results window. The goodness-of-fit statistics and diagnostics indicate the fit is fine:

```
    Estimation done by Maximum Likelihood (exact score)
    The databased used is USYCIMP.IN7
    The selection sample is: 51(2) - 85(4)  (T = 139, N = 1)
    The dependent variable Y is: Dp
    The model is:   Y = Level + Irregular
    Steady state. found

Log-Likelihood is 746.842 (-2 LogL = -1493.68).
Prediction error variance is 1.97066e-005
Summary statistics
                        Dp
T                   139.00
p                   1.0000
std.error           0.0044392
Normality           4.1757
H(46)               0.86535
DW                  1.9946
r(1)                0.00097185
q                   11.000
r(q)               -0.18240
Q(q,q-p)            7.1812
R^2                 0.62510

Variances of disturbances:
                Value     (q-ratio)
Level       2.84629e-006  (  0.2308)
Irregular   1.23330e-005  (  1.000)
```

To see the implications of the local level specification most clearly, go to Graphs/Model in the Documents window. These graphs can be reproduced by going to **Test**/Components graphics. Selecting 'Trend' in the box 'Plots with Y' and selecting 'Irregular', which are the defaults, and clicking on OK brings up a graph which shows the underlying rate of inflation. You can also move to the Forecast dialog. By accepting the default settings, you will see the underlying rate of inflation and its forecast; see Figure 4.2.

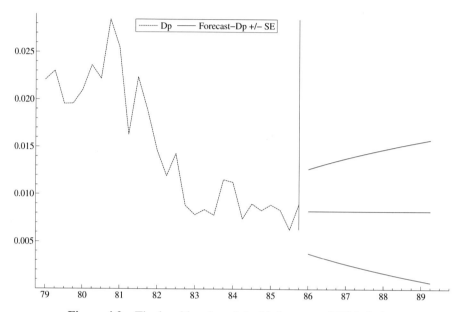

Figure 4.2 The local level model with forecast of US inflation.

The forecast function is a horizontal straight line. It projects the estimate of the current level into the future and this estimate is effectively constructed by putting exponentially declining weights on the past observations; see Harvey (1989). The RMSE shows how the uncertainty of the forecasts increases with the lead time. Again this is shown in Figure 4.2.

4.2.2 Statistical analysis of the local level model

The statistical specification consists of a random walk component to capture the underlying level, μ_t, plus a random, 'white noise', disturbance term, ϵ_t,

$$\begin{aligned} y_t &= \mu_t + \epsilon_t, & \epsilon_t &\sim \text{NID}(0, \sigma_\epsilon^2), & t = 1, \ldots, T, \\ \mu_t &= \mu_{t-1} + \eta_t, & \eta_t &\sim \text{NID}(0, \sigma_\eta^2), \end{aligned} \quad (4.1)$$

where η_t is the white noise disturbance driving the level. Both disturbances are normally distributed and independent of each other.

- When STAMP carries out estimation it is computing ML estimates of the variances, σ_η^2 and σ_ϵ^2. In the inflation example, these estimates are 2.85×10^{-6} and 1.23×10^{-5}. The relative variance, known here as the *signal to noise ratio*, is

$\sigma_\eta^2/\sigma_\epsilon^2$ or, if we are working in terms of standard deviations, $\sigma_\eta/\sigma_\epsilon$. It is given in parentheses after the variance estimate as the *q-ratio* and is given by 0.23.

- After estimation, STAMP runs a *Kalman filter* through the observations to estimate the *state*, μ_t. The estimate of the final state, μ_T, is available from the Test/More written output dialog by activating the option 'State and regression output' (in the section 'Print state vector'):

Coefficients of components in final state at period 85(4)

	Coefficient	RMSE	t-value	Prob
Level	0.00823	0.00216	3.80859	[0.00021]

For Dp it is 0.008, which translates to 3.2% per year ($4 \times 100 \times 0.008$), with a root mean square error (RMSE) of 0.8% ($4 \times 100 \times 0.002$). Thus the current annual rate of inflation, which is to be projected into the future, is about 3.2%.

- The **Test**/Components graphics dialog allows you to estimate the trend at all points in the sample using all the observations. This is known as *signal extraction* or *smoothing*. The *filtered* estimate of the trend can also be computed at all points in the sample if you wish. Unlike the smoothed estimate, the filtered estimate is based only on previous observations and the current observation. The *predicted* estimate is based only on previous observations and can be presented as well. The different estimates are presented in Figure 4.3. The graphs can be obtained from the **Test**/Components graphics by expanding the section 'Prediction, filtering and smoothing' (click on this title) and activating the options 'Predictive filtering' and 'Filtering'. Then press OK.

The inclusion of a random walk component in the model means that it is nonstationary. However, it is stationary in first differences, since

$$\Delta y_t = \eta_t + \epsilon_t - \epsilon_{t-1}, \quad t = 2, \ldots, T. \tag{4.2}$$

The OxMetrics Calculator and Graphics tools allow the user to create the differenced time series of Dp, that is DDp, and to produce graphs for the actual time series Dp and DDp and their associating correlograms. These graphs are presented in Figure 4.4 and they indicate quite clearly that differencing leads to a stationary series. (It can be shown to be equivalent to a first-order MA process.) However, while the transformation to stationarity is insightful, it should be stressed that this kind of analysis plays a much less prominent role in structural time series modelling than it does in the ARIMA methodology of Box and Jenkins (1970).

4.2.3 Local linear trend and smooth trend

Specifying both the level and the slope to be stochastic, which is the default, results in the local linear trend model. The forecast function is a straight line with an upward or downward slope, and both the level and the slope are constructed by putting more weight on the most recent observations.

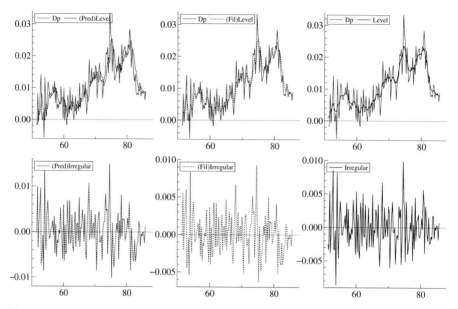

Figure 4.3 The local level model with different estimates of level and irregular components: predicted, filtered and smoothed estimates.

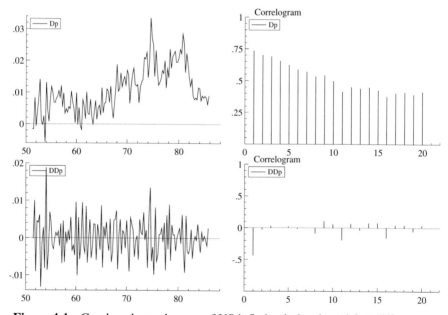

Figure 4.4 Graph and correlogram of US inflation in levels and first differences.

A variant of the local linear trend model induces a somewhat smoother trend by specifying 'Fixed Level' in the Select Components dialog. This is usually combined with an autoregressive component or cycle; see §4.4.4.

4.2.4 Statistical specification of the local linear trend model

A stochastic trend (local linear trend) model may be written

$$y_t = \mu_t + \epsilon_t, \quad \epsilon_t \sim \text{NID}(0, \sigma_\epsilon^2), \qquad (4.3)$$

where the trend component is specified as

$$\begin{aligned} \mu_t &= \mu_{t-1} + \beta_{t-1} + \eta_t, & \eta_t &\sim \text{NID}(0, \sigma_\eta^2), \\ \beta_t &= \beta_{t-1} + \zeta_t, & \zeta_t &\sim \text{NID}(0, \sigma_\zeta^2), \end{aligned} \qquad (4.4)$$

where μ_t is the level and β_t is the slope. An alternative way of representing the model for the trend μ_t is

$$\Delta \mu_t = \beta_{t-1} + \eta_t, \quad \Delta \beta_t = \zeta_t, \qquad (4.5)$$

where Δ is the difference operator ($\Delta = 1 - L$ with $Ly_t = y_{t-1}$). A more general version of the trend is given by

$$\Delta \mu_t = \beta_{t-1} + \eta_t, \quad \Delta^p \beta_t = \zeta_t, \qquad (4.6)$$

for $p = 1, 2, \ldots$ where $\Delta^p = (1 - L)^p$ with $L^p y_t = y_{t-p}$. When value $p > 1$ is taken, the slope component is more smooth. The estimated trend component from a trend model with $\sigma_\eta^2 = 0$ and $p > 1$ can be compared with a so-called Butterworth filter, see Gomez (2001).

The various statistical specifications of the trend in the Components dialog:

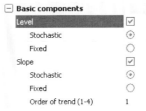

can now be defined precisely as follows.

- Level — determines whether level is present in model; if 'Level' is not selected, it is not present in the model and there can be no slope.
  - Stochastic — level specification contains η_t;
  - Fixed — level specification does not contain η_t; that is, σ_η^2 is set to zero;
- Slope — determines whether slope is present in the level specification; if 'Slope' is not selected, it is not present in the model.

- Stochastic — slope specification contains ζ_t;
- Fixed — slope specification does not contain ζ_t; that is, σ_ζ^2 is set to zero.
- Order of trend — the smoothness value p for the slope component in (4.6).

4.2.5 Specification of the trend

The default specification has $p = 1$ and both level and slope stochastic. The following are important special cases:

- *Local level* or *random walk plus noise* — the trend is a random walk and so we specify 'Level, stochastic' and 'Slope, not selected';
- *Local level with drift* — 'Level, stochastic' and 'Slope, fixed';
- *Smooth trend* — 'Level, fixed' and 'Slope, stochastic';
- *Butterworth trend* — 'Level, fixed', 'Slope, stochastic' and 'Order of trend p' with $p = 2, 3, 4$.

The specification of the trend can sometimes be based on prior knowledge or a plot of the data. If there is doubt, then estimate a general model. Tests of variances being zero have been developed by Nyblom and Makelainen (1983) and Harvey (2001). It is not unusual to find a variance going to zero. We then test if the corresponding element in the state is zero. In particular, if the slope variance (σ_ζ^2) is estimated to be zero, we may wish to test if the slope itself (β), which is now fixed, is also zero. This can be done in the Final state dialog by examining the '$t - value$'. Once restrictions have been imposed we need to check the serial correlation diagnostics.

Quite often more than one trend specification may be adequate in terms of diagnostics and goodness of fit. It is useful to examine the implications of different trends by looking at the smoothed components. It is also useful to check the forecast function. For example, a stochastic slope can sometimes be too sensitive to changes in the series, resulting in very unstable forecasts.

4.3 Seasonal

When appropriate, a seasonal component may be included in a structural time series model. If the data are inputted as quarterly or monthly, the number of seasons in a year, s, is known to STAMP, and a stochastic seasonal option is based on a trigonometric formulation. For daily observations, an intra-weekly pattern is assumed; that is, s is taken to be 7. The definition of the trigonometric form of stochastic seasonality is given in Chapter 9 and discussed at length in Harvey (1989). The important point to understand is that the seasonal pattern becomes deterministic if the seasonal variance parameter is set to zero.

4.3 Seasonal

4.3.1 Specifying and testing the seasonal component

There are three options for the seasonal component in the Components dialog:

```
Basic components
    Seasonal                    ☑
        Stochastic              ◉
        Fixed                   ○
        Select frequencies...   ☐
```

When a seasonal component is thought to be present, 'Seasonal' is selected. It may sometimes be appropriate to select 'Fixed' for the seasonal componen, perhaps because a previous attempt to fit the model indicated that this was the case. If the number of years is small, it may be reasonable to fix the seasonal as there is not enough data to allow a changing pattern to be estimated.

If it is believed that there is no seasonal component or if the series has been seasonally adjusted, the 'Seasonal' option may be de-activated. If such an assumption is inappropriate, the residuals will tend to show serial correlation, particularly at the seasonal lag, s.

Now select a typical seasonal series, for example, the logarithm of income, y, in the data file UKCYP.IN7. Select the default in Components. This consists of a stochastic level and slope, a stochastic trigonometric seasonal, and an irregular component. It is known as the *Basic Structural Model* (BSM). Estimate the model.

The diagnostic summary and estimated variances of disturbances are printed by default. The dialog Test/More written output provides more output. In this dialog, activate 'Anti-log analysis' and 'State and regression output'. The complete output is:

```
Log-Likelihood is 583.909 (-2 LogL = -1167.82).
Prediction error variance is 0.000351157
Summary statistics
                         y
T                   154.00
p                    3.0000
std.error            0.018739
Normality            5.6194
H(49)                0.54011
DW                   1.9924
r(1)                -0.0051813
q                   14.000
r(q)                -0.21686
Q(q,q-p)            31.238
Rs^2                 0.16946
```

```
Variances of disturbances:
                  Value        (q-ratio)
Level         0.000198515  (   1.000)
Slope         0.000000     (   0.0000)
Seasonal      1.85395e-006 (   0.009339)
Irregular     3.71636e-005 (   0.1872)
```

```
State vector anti-log analysis at period 93(2)
It is assumed that time series is in logs.
                              Value        Prob
Level (anti-log)          96250.09220   [0.00000]
Level (bias corrected)    96252.77834   [    -]
Slope (yearly %growth)        2.62979   [0.00000]
Seasonal chi2 test            9.05111   [0.02862]
Seasonal effects:
                Period     Value         Prob      %Effect
                   1       0.98435    [0.00941]   -1.56494
                   2       1.00009    [0.98824]    0.00908
                   3       1.00370    [0.57452]    0.37028
                   4       1.01206    [0.05865]    1.20585
```

The features of the output directly relevant to seasonal effects are as follows:

- The seasonal variance parameter is non-zero, indicating that there are changes in the seasonal pattern.
- The state has $s - 1$ elements to capture seasonality. These are not directly interpretable in the seasonal case, but STAMP transforms them into monthly or quarterly effects as appropriate. *Note that the seasonals sum to zero.*
- If the 'Anti-log analysis' option is marked, the seasonal effects are given as factors of proportionality by which to multiply the other components to get the systematic part of the series. Thus, in the case of UK income, the seasonal factors are 0.984, 1.000, 1.004 and 1.012. This indicates that in the fourth season, winter, the level of the trend needs to be multiplied by 1.012; in other words, incomes are, on average, 1.2% higher in quarter 4.
- The importance of the seasonal may be assessed by means of the 'Seasonal test' which is also printed. This tests the statistical significance of the seasonal pattern at the end of the sample period. It is asymptotically chi-square, here $\chi^2(3)$. The test may be misleading if the seasonal pattern is stochastic and it has changed a lot over the sample. Thus if the seasonals have become smaller, they may be insignificant at the end, but not at the beginning. In this case it is important to assess the test in conjunction with a full plot of the seasonal. This point is

relevant for UK income, where the seasonal effect becomes smaller over time as the economy becomes less dependent on seasonal factors.

The way in which the seasonal pattern has evolved over time is seen in the smoothed seasonal given by the Components graphics dialog. Selecting only 'Seasonal' (in the 'Select components' section) and 'individual seasonals' (in the 'Further plots' section) produces Figure 4.5. The bottom graph shows the yearly seasonal plots with the evolution of each of the seasonals over time.

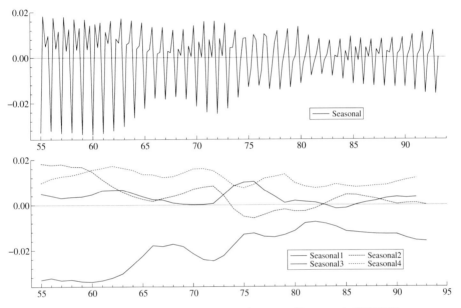

Figure 4.5 The estimated seasonal terms of income in UKCYP.

When forecasting is carried out, the latest seasonal pattern, as can be reported by State and regression output in the More written output dialog, is projected into the future.

```
Coefficients of components in final state at period 93(2)
             Coefficient       RMSE      t-value       Prob
Level           11.47471    0.00747   1535.91044   [0.00000]
Slope            0.00657    0.00114      5.76118   [0.00000]
Seasonal        -0.00595    0.00472     -1.25957   [0.20979]
Seasonal 2       0.00973    0.00481      2.02343   [0.04482]
Seasonal 3       0.00604    0.00356      1.69492   [0.09218]
```

4.3.2 Seasonal adjustment

If a model has been fitted with a seasonal component, optimal model-based seasonal adjustment may be carried out within the Components graphics dialog by marking 'Seasonally adjusted Y' in the 'Further plots' section. This option subtracts the smoothed seasonal component from the full series. It can be stored for future use. A more detailed discussion of seasonal adjustment can be found in Chapter 7.

If the data has already been seasonally adjusted by another method, the 'No seasonal' option will normally be chosen. However, not all seasonal adjustment procedures succeed in removing seasonality. If this is the case, the residuals will tend to show serial correlation, particularly at the seasonal lag, s. A more thorough check can be carried out by including a seasonal component and seeing if it appears to be significant.

4.4 Cycle

A deterministic cycle is a sine-cosine wave with a given period. A stochastic cycle is constructed by shocking it with disturbances and introducing a damping factor. Such stochastic cycles are capable of modelling the kind of pseudo-cyclical behaviour which is characteristic of many time series, particularly economic and social ones. A deterministic cycle emerges as a limiting case. In many areas, such as meteorology, the question of whether cycles should be deterministic is an open one and so having such cycles as a special case of a more general model is very useful.

STAMP allows the inclusion of up to three stochastic cycles in a model. These are specified in the Select components dialog by marking boxes. If you have less than three cycles, which boxes you mark is a matter of convenience, determined by the default settings.

4.4.1 A simple cycle plus noise model

The RAINBRAZ.IN7 file consists of a single series, RainFort, which is annual data on the number of centimetres of rainfall in Fortaleza, a town in north-east Brazil. The series goes up to 1992 but we will conduct the initial analysis and modelling on the data up to 1984 only.

It is instructive to carry out a preliminary analysis on the data using the OxMetrics options. Select the Graphics dialog from the **Model** menu in OxMetrics. Select the only variable in the 'Database' listbox and press the 'All plot types >' button. In the section 'Actual series', select 'Sample' (double-click on 'full sample' or press button

'...') and change the end of the sample to 1984. Then click on the graphical button 'Actual series' in the left-side panel and press 'Plot'. To view the correlogram, move to the time series section by clicking on 'Time-series properties' at the top-panel of the Graphics dialog and click on the graphical button 'Autocorrelation function (ACF)' (in left-side panel) and press 'Plot'. To view the spectrum, click on 'Spectrum' (more below) and press 'Plot'. The histogram of the series can be selected from the section 'Distribution'. Cancel the dialog to view the graphs as shown in Figure 4.6. We would expect a rainfall series to have a constant mean and this seems to be borne out by a graph of the series. Although the individual autocorrelations are quite small, the correlogram shows evidence of a cycle buried within the noise. The same message appears in the estimated spectrum, but more clearly. On this graph, the period is given by dividing 2 by the scaled frequency, on the horizontal axis. Thus there appears to be a cycle with a period of around 12 or 13 years. Re-estimating the spectrum with the window implied by 50 lags, indicates the possibility of a second cycle of around 25 years. There is also a smaller peak in the spectrum at around four years.

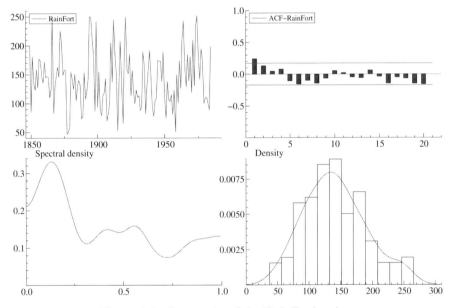

Figure 4.6 Summaries of the 'RainFort' series.

For the moment we will assume a single cycle. The possibility of such a cycle leads us to formulate a model consisting of a 'Level, Fixed', an 'Irregular' and a 'Cycle'. The 'Slope' component is not selected. To select the cycle, open the section 'Cycle(s)' in the Select components dialog and select 'Cycle medium'. This choice is convenient since the default is for the estimation procedure to start off with a period of 10. Set the end of the sample to 1984 in the Estimation dialog.

The most easily interpretable part of the output is the graph of the 'Cycles and ARs' and the 'Trend plus Cycles plus ARs' in the Test/Component graphics dialog. It can be seen in Figure 4.7 that the cycle is somewhat irregular in period and amplitude, and is dominated by the irregular component. The forecasts show that the cycle damps down towards zero as the lead time increases.

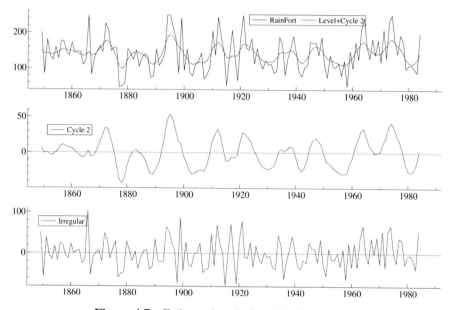

Figure 4.7 Estimated cycle for 'RainFort' series.

More precise information about the fitted cycle can be found in the More written output dialog. Selecting 'Parameters by component' prints for the cycle and irregular components:

```
Parameters in Cycle

Variance                    726.30
Period                      14.748
Frequency                   0.42603
Damping factor              0.85031

Parameters in Irregular

Variance                    1582.1
```

The parameters, which are defined formally in the next subsection, are as follows:

- A variance parameter which is responsible for making the cycle stochastic, σ_κ^2;
- A period (in years), $2\pi/\lambda_c$
- A frequency (in radians), λ_c.
- A damping factor, ρ;

The information in the frequency is more usefully presented in terms of the period (this is 2π divided by the frequency λ_c). The output shows this to be just under 15 years.

The variance of the cycle itself, as opposed to the variance of the disturbance term generating it, is also shown. The sum of this variance and the variance of the irregular should, in theory, be equal to the variance of the observed variable.

4.4.2 Statistical specification

The statistical specification of a cycle, ψ_t, is as follows

$$\begin{bmatrix} \psi_t \\ \psi_t^* \end{bmatrix} = \rho \begin{bmatrix} \cos\lambda_c & \sin\lambda_c \\ -\sin\lambda_c & \cos\lambda_c \end{bmatrix} \begin{bmatrix} \psi_{t-1} \\ \psi_{t-1}^* \end{bmatrix} + \begin{bmatrix} \kappa_t \\ \kappa_t^* \end{bmatrix}, \quad t=1,\ldots,T \quad (4.7)$$

where λ_c is the frequency, in radians, in the range $0 < \lambda_c < \pi$, κ_t and κ_t^* are two mutually uncorrelated white noise disturbances with zero means and common variance σ_κ^2, and ρ is a *damping factor*. Note that the *period* is $2\pi/\lambda_c$. The stochastic cycle becomes a first-order autoregressive process if λ_c is 0 or π.

In estimating the model, the variance of the cycle itself, σ_ψ^2, rather than σ_κ^2, is taken to be the fixed parameter. Since $\sigma_\kappa^2 = (1-\rho^2)\sigma_\psi^2$, it follows that $\sigma_\kappa^2 \to 0$ as $\rho \to 1$ and (4.7) reduces to the deterministic, but stationary, cycle (4.8)

$$\psi_t = \psi_0 \cos\lambda_c t + \psi_0^* \sin\lambda_c t, \quad t=1,\ldots,T, \quad (4.8)$$

where ψ_0 and ψ_0^* are uncorrelated random variables with zero mean and common variance σ_ψ^2.

The cycle plus noise model of the previous subsection may be written as

$$y_t = \mu + \psi_t + \epsilon_t, \quad (4.9)$$

where μ is the mean of the series and ϵ_t is a white noise term which is uncorrelated with ψ_t and has variance σ_ϵ^2. The decomposition of the variance of y_t may be written as:

$$\sigma_y^2 = \sigma_\psi^2 + \sigma_\epsilon^2. \quad (4.10)$$

4.4.3 Higher order cycles

The cycle component specification can be generalised as proposed by Harvey and Trimbur (2003). Smoother cycle processes can be specified as $\psi_t = \psi_t^{(k)}$ where

$$\begin{bmatrix} \psi_t^{(j)} \\ \psi_t^{*(j)} \end{bmatrix} = \rho \begin{bmatrix} \cos\lambda & \sin\lambda \\ -\sin\lambda & \cos\lambda \end{bmatrix} \begin{bmatrix} \psi_{t-1}^{(j)} \\ \psi_{t-1}^{*(j)} \end{bmatrix} + \begin{bmatrix} \psi_t^{(j-1)} \\ \psi_t^{*(j-1)} \end{bmatrix}, \quad (4.11)$$

for $j = 1,\ldots,k$, and where $\kappa_t = \psi_t^{(0)}$ and $\kappa_t^* = \psi_t^{*(0)}$ are two mutually uncorrelated white noise disturbances with zero means and common variance σ_κ^2 and for

$t = 1, \ldots, n$. A higher value for k leads to more pronounced cut-offs of the band-pass gain function at both ends of the range of business cycle frequencies centered at λ. For example, the model-based filter with $k = 6$ leads to a similar gain function as the one of Baxter and King (1999). Different variations within this class of generalised cycles are discussed in Harvey and Trimbur (2003) who refer to specification (4.11) as the balanced cycle model. They prefer this specification since the time-domain properties of cycle ψ_t are more straightforward to derive and it tends to give a slightly better fit to a selection of U.S. economic time series.

The value of k determines the smoothness of the cycle. In the case $k = 1$, the generalised cycle reduces to (4.7). The value of k, the 'Order of cycle', can be set in the 'Cycle(s)' section of the Select components dialog and has possible values $1, 2, 3, 4$. Illustrations of the higher order cycle are given in §7.1.

4.4.4 Trend plus cycle

The series on US GNP is the y series in USYCIMP. It yields an important example of a trend plus cycle model.

In the Select components de-select 'Seasonal' component in the 'Basic components' section and 'Cycle short' in the 'Cycle(s)' section. Also select 'Level, Fixed' so as to give a smooth trend component. *The use of a smooth trend with a cycle often leads to a more attractive decomposition.* Estimate the model. The period of the cycle is about five years, corresponding to a plausible business cycle. The forecast plot, which was displayed earlier in Figure 1.3, is even more informative. It turns out that the peaks and troughs follow the movements charted by the National Bureau of Economic Research (NBER).

Other features of the results include the fact that the irregular variance has been estimated to be zero. Thus it can be — and in fact effectively has been — dropped from the model.

4.4.5 Multiple cycles

The model in (4.9) can be extended so as to include several cycles at different frequencies. Returning to RAINBRAZ and marking cycles 'Cycle medium' and 'Cycle long' (in addition to fixed level and no slope) gives the following results with data up to 1984:

```
UC( 1) Modelling RainFort by Maximum Likelihood (using RAINBRAZ.IN7)
    The selection sample is: 1849 - 1984
    The model is:   Y = Level + Irregular + Cycle 2 + Cycle 3

Log-Likelihood is -511.383 (-2 LogL = 1022.77).
Prediction error variance is 1740.01

Summary statistics
 std.error          41.713
 Normality           0.69437
```

```
H(45)            0.91860
r(1)             0.035794
r(10)            0.15912
DW               1.8722
Q(10,7)          7.9241
R^2              0.24116

Variances of disturbances.

Component                    Value      (q-ratio)
Level                      0.00000    ( 0.0000)
Cycle                      0.00067795 ( 0.0000)
Cycle 2                    0.00066725 ( 0.0000)
Irregular                  1748.8     ( 1.0000)

Parameters in Cycle

Variance                      338.97
Period                         12.958
Frequency                       0.48489
Damping factor                  1.0000

Parameters in Cycle 2

Variance                      224.96
Period                         24.587
Frequency                       0.25554
Damping factor                  1.0000
```

State vector analysis at period 1984
- level is 143.356 with stand.err 3.59988.
- amplitude of Cycle 2 is 25.0662
- amplitude of Cycle 3 is 20.0908

Using more mathematical notation, this can be summarized as:

$$\tilde{\sigma}^2_{\psi 1} = 339, \quad \tilde{\rho} = 1, \quad \tilde{\lambda}_c = .485 (\text{period} = 12.96), \quad \tilde{\sigma} = 41.7,$$
$$\tilde{\sigma}^2_{\psi 2} = 225, \quad \tilde{\rho} = 1, \quad \tilde{\lambda}_c = .256 (\text{period} = 24.59), \quad Q(10,7) = 7.9, \quad N = .69,$$
$$\tilde{\sigma}^2_\epsilon = 1749, \qquad\qquad\qquad\qquad\qquad\qquad\qquad R^2 = .24, \quad H = .92.$$

Thus estimation of the two-cycle model gives two deterministic cycles. The sum of the variance components is 2313; this is close to the series variance of 2293. The diagnostics are satisfactory and if a third cycle is included in the model, it is either very small or disappears completely, depending on the starting valuesused. The results are similar to those reported in Kane and Trivedi (1986) and Morrettin, Mesquita and Rocha (1985) where the authors used a regression model with the two frequencies determined from prior ideas or an inspection of the periodogram.

The two-cycle model appears to be stable and its predictive performance is rather good. Re-estimating the model using observations up to 1992 changes the parameter estimates very little and extrapolating from 1972 would have clearly predicted the droughts of the early 80s and 90s. In the stochastic cycle model the cycle in the forecast function tends to die away after a few years.

4.5 Autoregression

First-order and second-order autoregressive — AR(1) and AR(2) — components can be included in a structural time series model by marking the 'AR(1)' and/or 'AR(2)' box in the Select components dialog. The process is constrained to be stationary; that is, the AR coefficients are restricted to represent a stationary process. If this were not the case there would be a risk of them being confounded with the random walk component in the trend. As noted in the previous section, the stochastic cycle becomes an AR(1) if λ_c is 0 or π, but this possibility is avoided by constraining λ_c to be strictly between 0 and π. Obviously if it ends up being close, reformulating the cyclical component as an AR is probably appropriate.

If the series itself is a pure AR, of any order, rather than an AR plus other stochastic components, it may be estimated by OLS by using the 'Lag' option in the Formulate dialog to create lagged values of the dependent variable. As an example, consider the inflation series in USYCIMP, Dp, used in §4.2, and formulate a model for the first differences with three lags. To include the lags, you select in the 'Lags' section the option 'Lag 0 to' (instead of the default 'None') and type 3 in the next box. Then double-click on 'Dp' in the 'Database' selection or press on << after selecting 'Dp'. As a result, the actual series 'Dp' appears as a Y variable and the three lags are included in the model as explanatory variables.

After estimation, the AR coefficients are printed using the 'Parameters by component' option in Test/More written output. *A useful theoretical exercise is to expand the reduced form MA(1) process obtained from equation (4.2) as an infinite AR and compare the implied coefficients with those of the fitted AR(3); see Harvey (1993, p. 133).*

4.6 Exercises

(1) Fit a local linear trend model to the price level, p, in USYCIMP. Compare the results with the local level model.
(2) The PURSE file contains observations on the number of purses (handbags) snatched in the Hyde Park area of Chicago. Fit a local level model, trying various transformations, such as logarithms, square roots and other powers, for example 2/3. Which is best? What happens if you include a slope in the model?
(3) The series in USYCIMP have been seasonally adjusted by the US Census Bureau's X-11 program. Fit a local level model to Dp. What do you conclude about the effectiveness of X-11 in this case?

(4) Estimate the RAINBRAZ series with a stochastic level. What do you conclude?

- In the article by Newton, North and Crowley (1991), it is suggested that the global ice volume series, contained in ICEVOL, contains cycles of around 100,000, 41,000 and 22,000 years, but that these may not be strictly periodic. Bearing in mind that the points are 2,000 years apart, estimate a model with these three cycles and make predictions for the next 100,000 years.

(5) Fit a basic structural model to ofuGASl in ENERGY. Compare the fit when the seasonal is constrained to be fixed. Do the diagnostics indicate a misspecification?

(6) Seasonally adjust the series in the previous question using the BSM. Are there any seasonal effects remaining?

(7) Estimate the rainfall series in RAINBRAZ with three cycles. How does the predictive performance compare with the two-cycle model?

Chapter 5

Tutorial on Interventions and Explanatory Variables

5.1 Interventions

Intervention variables are dummy (or indicator) variables which are used to take account of outlying observations and structural breaks. These data irregularities are usually thought of as arising from a specific event, for example a strike in the case of an outlier or a change in policy in the case of a structural break.

An *outlier* can be thought of as an unusually large value of the *irregular* disturbance at a particular time. It can be captured by an *impulse* intervention variable which takes the value one at the time of the outlier and zero elsewhere.

A *structural break* in which the level of the series shifts up or down is modelled by a *step* intervention variable which is zero before the event and one after. Alternatively it can be modelled in exactly the same way by adding an outlying intervention to the *level* equation. In other words the break is identified with an unusually large value of the level disturbance.

A *structural break in the slope* can be modelled by a *staircase* intervention which is a trend variable taking the values, 1, 2, 3, . . ., starting in the period after the break. It can be thought of as arising from a large value of the *slope* disturbance.

The easiest way to include intervention variables in the model is by by choosing the option 'Select interventions automatically' using the appropriate radio-button in section 'Options' of the Select a model dialog. The program selects the appropriate outliers and breaks in the time series using an intelligent procedure and described below. Alternatively, the interventions can be inputted manually by accessing the Intervention dialog. This is done from the Select components dialog by pressing the radio-button 'Select interventions manually' in section 'Options'. Level (step) and slope (staircase) interventions formed in this way have their effects included in the trend. The selected interventions can be saved for future use. Alternatively, intervention variables can be created as explanatory variables by entering them directly in the database or by using the Calculator. Interventions formed in this way can be saved in the database for future use.

5.1 Interventions

The next subsection will use the database NILE.IN7.

5.1.1 Modelling

The series contained in NILE is of the volume of the flow of the river, in cubic metres ($\times 10^8$) in the years from 1871 to 1970. We will model this as a fixed level plus a cycle plus an irregular with a structural break in 1899, corresponding to the building of a dam at Aswan and an outlier in 1913. The route by which we might have arrived at such a model is explored in the next section. Our methodology is very easy to implement — it is instructive to compare it with the ARIMA approach described in Balke (1993).

Having loaded the data, go to the Formulate dialog in the Model menu (Alt+y), mark 'Nile', press << and, finally, press OK to move to Select components. Enter the specification suggested above in the standard way by selecting 'Level, Fixed', deselecting 'Slope' and selecting 'Medium cycle' in the 'Cycle(s)' section. To access the Intervention dialog, mark 'Select interventions manually' in the 'Options' section of the Select components dialog and press OK:

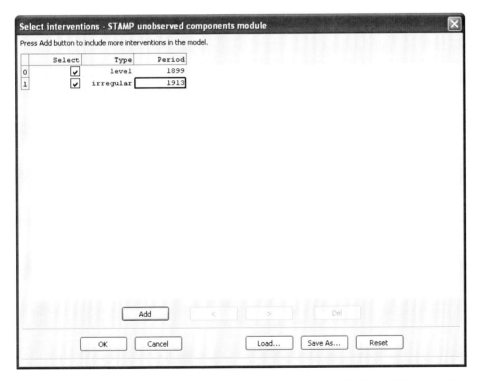

This dialog has one entry as a default. Entries can be added by clicking on the 'Add' button. Put in the break by selecting the entry in the 'Select' column, select 'Level' in the drop-box of the second column 'Type' and input the year 1899 in the

'Period' column. The level break for 1899 has become a part of the trend component. The outlier in 1913 can be inputted in the same way but selecting the 'Irregular' in the 'Type' column and 1913 in the 'Period' column. (It will be also useful to note that you can clear, from the present model, the interventions by deselecting the intervention in the 'Select' column.) Once the interventions have been added, click on OK and estimate the model.

A part of the default output is given by:

```
Regression effects in final state at time 1970
                   Coefficient        RMSE        t-value        Prob
Level break 1899. 1  -243.28229     30.02331     -8.10311      [0.00000]
Outlier 1913. 1      -377.86556    119.51527     -3.16165      [0.00209]
```

As can be seen, both interventions are statistically significant when judged against t-distributions. The two-sided probability values are shown in square brackets. Although a t-distribution is, strictly speaking, only valid if the (relative) variances are known, it provides a good guide in practice.

Now move to the **Test**/Component graphics dialog and keep the default sets of graphs. An inspection of 'Trend plus Regression effects' and 'Fixed intervention effects' will show a clear shift in the level after 1899, as the step intervention has been absorbed into the trend component.

Finally enter the **Test**/Forecasting dialog and select 'Signal' and 'Trend plus Regression effects'. Note that the forecast of the signal continues at the lower level, incorporating the effect of the intervention.

5.1.2 Detection using auxiliary residuals

Outliers and structural breaks can often be detected simply by looking at a graph of the series. When a strong seasonal pattern is present, it may be better to seasonally adjust the data first. If you look at the seasonally adjusted ofuCOAL1 series, for example, possible breaks and outliers can be seen which are less apparent from an inspection of the original series.

Although looking at a graph is always helpful, it is often not possible to detect outliers and structural breaks or to distinguish between them. This is where the auxiliary residuals, due to Harvey and Koopman (1992), are useful. *The auxiliary residuals are smoothed estimates of the irregular and level disturbances* and although they are neither serially uncorrelated nor uncorrelated with each other, they play a valuable role in that they go some way towards separating out pieces of information which are mixed up together in the innovation residuals. It is possible to construct estimates of other disturbances, such as the one driving the slope, but experience suggests that these are less useful.

We illustrate the use of the auxiliary residuals by looking at how we might have gone about specifying the model for the Nile flow estimated in the previous subsection.

5.1 Interventions

If you still have the NILE data loaded, go to **Model**/Select components and specify a model with no slope, but allow the level to remain stochastic. Do not specify a cycle and remember to take out the interventions by accessing the Interventions dialog and deselecting them. The model is therefore a simple random walk plus noise. This might well be a plausible initial specification.

Having estimated the model, examine the trend in **Test**/Components graphics. It can be seen how the level adapts to the change in 1899 and picks up the cycle. The Auxiliary residuals graphics dialog, which can be found in the **Test** menu, allows the inspection of irregular, level and slope residuals:

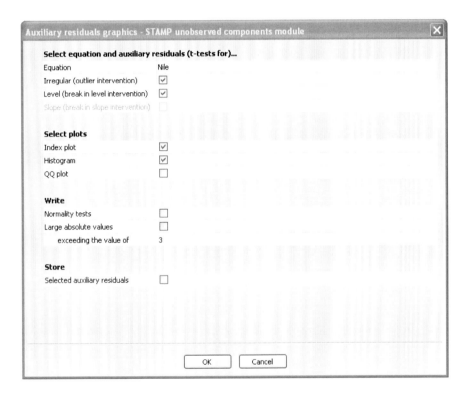

They provide an even clearer pointer to the structural break in the Nile series. Figure 5.1 shows a pronounced dip in the levels residual. The outlier in 1913 shows up clearly in the irregular. The normality statistics are not significant, but this only serves to stress the importance of looking at graphical output rather than relying solely on test statistics.

Once interventions are put into a model (re-activate them in the Select interventions dialog) with a fixed level, a cyclical pattern can be seen in the residual autocorrelation function, though the first-order autocorrelation, $r(1)$, is not significant.

72   Chapter 5 Tutorial on Interventions and Explanatory Variables

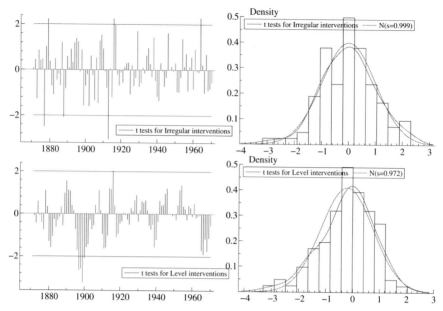

Figure 5.1 Auxiliary residuals for random walk + noise model of the Nile.

5.1.3 Automatic outlier and break detection

The inspection of auxiliary residuals is useful for the analysis and modelling of time series but can be time-consuming when the primary interest of the analysis is not focussed on outliers and breaks. The option 'Select interventions automatically' in the **Model**/Select components dialog allows the program to detect significant outliers and breaks and to include them in the model automatically without the interference of the user.

The strategy of finding outliers and breaks is based on the auxiliary resiudals and consists of the following two steps:

(1) The selected model is estimated (this model may include explanatory variables and imposed interventions, see discussion below). The auxiliary residuals are computed and the time periods of residuals that have absolute values exceeding 2.3 (for irregular), 2.5 (for level residual) or 3.0 (for slope residual) are recorded. Adjacent time periods for level and slope residuals are removed within a distance of upto 3 (for level) or 4 (for slope) periods from the largest residual. This addresses the issue that level and slope residuals are serially correlated over time, see Harvey and Koopman (1992).

(2) The model with the inclusion of interventions detected in Step 1 is re-estimated. The estimated interventions with t-values that do not exceed the value 3 in absolute values are de-selected as intervention variables from the model. The output reported in STAMP is based on this model.

5.1 Interventions

After Step 2, the model with some intervention variables excluded is not re-estimated. It should be re-estimated since the model specification has changed. However, since the removed interventions are not estimated as significant, the estimates of the remaining variables (including disturbance variances of the components and regression coefficients) may not be affected by much when these non-significant effects are removed from the model specification. It is nevertheless advisable, after re-inspecting the resulting model (using the various diagnostic statistics available in STAMP) and after re-considering the automatically selected set of interventions, to re-estimate the model. The automatic procedure will be illustrated using the NILE data.

Go to the **Model**/Select components dialog and specify a model with 'stochastic level', no 'slope', do not include a cycle, and activate 'Select interventions automatically' by pressing the appropriate radio-button. By pressing OK and accepting the settings in the Estimate dialog, the program carries out the two-step procedure as described above. In the results window, we find the output

```
Variances of disturbances:
                 Value     (q-ratio)
Level          0.000000  (   0.0000)
Irregular      14124.7   (   1.000)

State vector analysis at period 1970
           Value      Prob
Level   1097.75000  [0.00000]

Regression effects in final state at time 1970
                   Coefficient     RMSE       t-value      Prob
Level break 1899. 1  -242.22887    26.52156   -9.13328  [0.00000]
Outlier 1913. 1      -399.52113   119.68141   -3.33821  [0.00120]
```

The automatic procedure confirms the findings in the previous section. To check whether other outliers and breaks have also been considered in Step 1 of the procedure, proceed to the Select components dialog and accept the default settings which include the activated option of 'Select interventions manually'. This last option is activated as a default when intervention variables are part of the current model. In the Select interventions dialog, the level break of 1899 and the outlier of 1913 are activated while the outlier effect for 1877 is de-activated.

Chapter 5 Tutorial on Interventions and Explanatory Variables

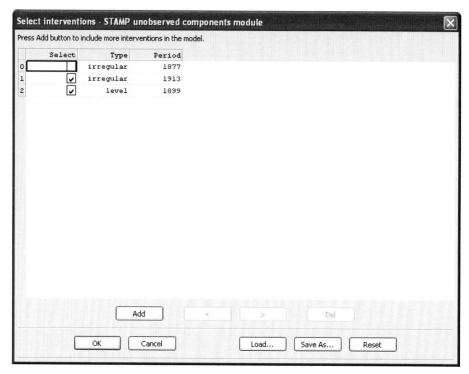

The outlier for 1877 was detected as a potential intervention variable in Step 1 of the procedure while it was dismissed as an outlier in Step 2 (its estimated t-value did not exceed the value of 3). The model may now be re-estimated with the two activated intervention variables included in the model. The estimation output has changed slightly since the variances of the components are re-estimated without the outlier intervention variable for 1877.

```
Variances of disturbances:
                   Value     (q-ratio)
Level           0.000000   (   0.0000)
Irregular       14845.9    (   1.000)

State vector analysis at period 1970
              Value       Prob
Level      1097.75000  [0.00000]

Regression effects in final state at time 1970
                    Coefficient      RMSE     t-value     Prob
Level break 1899. 1   -242.22887   27.19026   -8.90866  [0.00000]
Outlier 1913. 1       -399.52113  122.69901   -3.25611  [0.00156]
```

The estimated irregular variance has increased to the value of 14845.9 since the intervention variable for the outlier of 1877 has been excluded and has become part of the irregular component.

Finally, the automatic outlier and break detection procedure may be repeated with imposing pre-selected intervention variables in the model. Return to the Select components dialog and change the default setting from 'Select interventions manually' to 'Select interventions automatically'. By pressing OK, the Select interventions dialog appears and intervention variables can be added, deleted, selected or deselected. The selected interventions will be imposed in the model during the automatic outlier and break detection procedure described. In other words, they are regarded as a set of explanatory variables that remain in the model during the two-step procedure. In case of the NILE data, after repeated applications of the automatic outlier and break detection procedure, no new outliers and breaks will be detected.

In case of noisy data-sets, a long list of selected and non-selected intervention variables may appear in the Select interventions dialog. Although it is useful that the program keeps a record of all considered interventions in the past, it can also be useful to have an option that deletes all (selected and non-selected) intervention variables. This option is simply the 'Select interventions none' in the Select components dialog. When the model is estimated with this option activated, all intervention variables will be deleted from memory.

5.1.4 Specification of more complex interventions

The interventions considered so far have been of a relatively simple form. We now examine a somewhat more complicated case in which:

(1) the form of the intervention variable is such that it must be created as an explanatory variable; and
(2) the effect of the intervention in the periods immediately after the occurrence of the intervention is not known with certainty and so the form adopted must be subjected to *post-intervention diagnostics*.

The study by Harvey and Durbin (1986) on the effect of the seat belt law in Great Britain provides a good illustration of the above issues. Load the SEATBELT data. The intervention variable, 'law', is already included. It is basically a step intervention starting from February 1983, but it takes the value 0.18 in January because of increased seat belt usage in anticipation of the coming of the law on the last day of the month. It is a useful exercise to recreate this intervention. Go to **Model**/Calculator of OxMetrics, Alt+c. Press 'Dummy', and type 83 and 2, respectively, in the two edit boxes after 'Sample start'. Subsequently, press OK and, in the Calculator dialog, press '='. Give the variable a name, say 'law1', and press OK. Then leave the Calculator. In the database of SEATBELT, go to the field corresponding to row '83-1' and column 'law1' and, by double-clicking it or by selecting it and pressing ↩, you can edit the field and type

0.18. You now have an intervention exactly like 'law', but since you don't need it, go back to the calculator and delete it!

Now formulate a model for 'Drivers' and include 'law' as an explanatory variable. Use a BSM - the default in Components - or do not select 'Slope' and select 'Seasonal, Fixed' and then Estimate the model. The coefficient of the intervention appears in More written output, provided the 'State and Regression output' box is marked. The coefficient (with no slope and fixed seasonals) is -0.26, indicating a fall of 23%, that is $1 - \exp(-.26) = .23$, in car drivers killed and seriously injured after the introduction of the law.

To check the specification of the intervention variable, go to the **Test**/Prediction graphics dialog, keep the default settings and press OK.

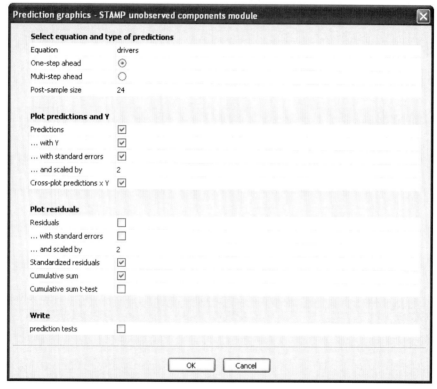

The plots, shown in Figure 5.2, show the predictive performance of the model after the intervention. It appears to be good, and the *post-intervention predictive test*, given here as the *Failure Chi2 test*, is not significant.

```
Prediction analysis for 24 post-sample predictions
(with 1 missing values).
               error   stand.err  residual   cusum    sqrsum
    83(1)      .NaN    0.07546    0.0000    0.0000   0.0000
```

```
83(2)        0.2607      0.4041      0.6452    0.6452    0.4163
83(3)        0.1212      0.09177     1.321     1.966     2.161
...
84(10)       0.07777     0.07526     1.033     6.568    18.91
84(11)       0.04815     0.07526     0.6399    7.208    19.32
84(12)      -0.007602    0.07526    -0.1010    7.107    19.33

Post-sample predictive tests.
Failure Chi2( 23) test is   19.3299  [0.6819]
Cusum t( 23)      test is    1.4820  [0.1519]

Post-sample prediction statistics.
Sum of 23 absolute prediction errors is 1.58736
Sum of 23 squared  prediction errors is 0.181874
Sum of 23 absolute prediction resids is 17.8311
Sum of 23 squared  prediction resids is 19.3299
```

Although the 'post-sample size' is set to 24 in the Prediction graphics dialog, only 23 residuals can be used for the test statistics. The residual of 83(1), with value .NaN, is lost since the regression effect 'law' takes effect from period 83(1) onwards and this is the first period in the post-sample of length 24. The observation at 83(1) is used to identify the regression effect. The uncertainty of the residuals in the period just after 83(1) is apparent from the higher standard errors and from the confidence interval in the prediction graph of Figure 5.2.

To carry out a post-intervention test over a shorter period, return to the **Test/Prediction** graphics dialog and proceed as follows. Change value 'Post-sample size' from 24 to 6 and mark 'prediction tests' in the 'Write' section. The Chow failure test statistic is still insignificant.

```
Post-sample predictive tests.
Failure Chi2( 6) test is   3.4923  [0.7450]
Cusum t( 6)      test is   1.1186  [0.3061]
```

5.2 Explanatory variables

Including explanatory variables in a structural time series model results in a mixture of time series and regression. Indeed classical regression emerges as a special case in which there are no stochastic components, apart from a single random disturbance term. STAMP will estimate classical regression models quite happily, but its real power comes in combining regression with unobserved components.

The combination of unobserved stochastic components with explanatory variables opens up a wide range of possibilities for dynamic modelling.

78 Chapter 5 Tutorial on Interventions and Explanatory Variables

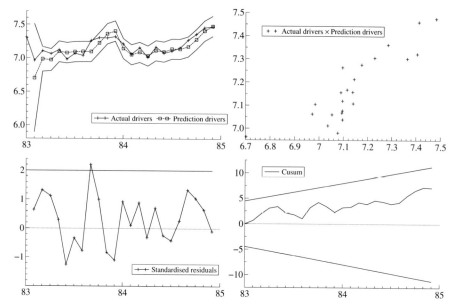

Figure 5.2 Predictive testing for the Drivers series.

5.2.1 Stochastic trend component

The SPIRIT data consists of annual observations, from 1870 to 1938, on the logarithms of three variables, the *per capita* consumption of spirits in the UK, *per capita* income, and the relative price of spirits. The data set is a famous one, having been used as a test bed for the Durbin–Watson statistic in 1951.

The series may be graphed in OxMetrics. Income and price are explanatory variables and the standard econometric approach is to estimate a regression model with a linear or quadratic time trend and an AR(1) disturbance; see Fuller (1996, p. 522). The structural time series modelling approach is simply to use a stochastic trend with the explanatory variables. The aim of the stochastic trend is to pick up changes in tastes and habits which cannot be measured explicitly.

Enter the Formulate dialog. Select the three variables by clicking and press << or by double-clicking. The first variable to be selected is 'Spirits' and it will be marked with 'Y' in the 'Selection' list box. The other two variables 'Income' and 'Price' will not be marked in the 'Selction' list box. This is as it should be. Press OK to enter the Select components dialog which is set for a stochastic level and slope with an irregular. There is no seasonal — obviously since the observations are annual. Since the model includes explanatory variables, the option 'Set regression coefficients' is marked by default. It allows you to enter the Regression coefficients dialog and to change the specification of the various regression coefficients. For example, coefficients can be specified as time-varying and modelled as a random walk process. For this illustration

5.2 Explanatory variables

we concentrate on fixed regression coefficients and therefore accept the default settings in this dialog and press OK. This brings up the Estimation dialog. Change the period of 'Estimation ends at' from 1938 to 1930. Press OK and once estimation is complete, the standard output reads:

```
The databased used is SPIRIT.IN7
The selection sample is: 1870 - 1930 (T = 61, N = 1)
The dependent variable Y is: spirits
The model is:  Y = Trend + Irregular + Explanatory vars
```

Log-Likelihood is 217.767 (-2 LogL = -435.535).
Prediction error variance is 0.000486587
Summary statistics

	spirits
T	61.000
p	2.0000
std.error	0.022059
Normality	7.9501
H(19)	2.2760
DW	2.0779
r(1)	-0.048381
q	8.0000
r(q)	-0.17059
Q(q,q-p)	4.1047
Rd^2	0.72830

Variances of disturbances:

	Value	(q-ratio)
Level	9.04323e-005	(0.6070)
Slope	3.55317e-005	(0.2385)
Irregular	0.000148994	(1.000)

State vector analysis at period 1930

	Value	Prob
Level	2.23692	[0.00000]
Slope	-0.01538	[0.12573]

Regression effects in final state at time 1930

	Coefficient	RMSE	t-value	Prob
price	-0.94969	0.07094	-13.38671	[0.00000]
income	0.69526	0.13199	5.26766	[0.00000]

All appears to be well, although the normality statistic (which measures the departure of third and fourth moment from their values expected under normality) is somewhat on the high side. Here H is a heteroskedasticity statistic, $r(1)$ and $r(8)$ are the serial correlation coefficients at the 1st and 8th lag. DW is the Durbin–Watson statistic, while $Q(8, 6)$ is the Box–Ljung statistic using 8 lags. $Q(8, 6)$ should have a χ_6^2 distribution under the null distribution. Finally R_d^2 is a measure of goodness-of-fit, computed on the differences of the spirit data set. All of these statistics will be discussed in more detail in Chapter 10.

We now turn to the **Test** menu. The first option is 'Variances' in the More written output dialog:

```
Variances of disturbances:
                Value           (q-ratio)
Level        9.04323e-005   (  0.6070)
Slope        3.55317e-005   (  0.2385)
Irregular    0.000148994    (  1.000)

Standard deviations of disturbances:
                Value           (q-ratio)
Level        0.00950959     (  0.7791)
Slope        0.00596084     (  0.4883)
Irregular    0.0122063      (  1.000)
```

The output shows the estimates of both the variances and the standard deviations of the disturbances driving the level and slope and the irregular disturbance. The first two are non-zero indicating a stochastic trend. However, the most useful information on the nature of this trend appears when we examine a plot of it later in Components.

Part of the standard output are the estimates of the coefficients of the explanatory variables. *These can be interpreted in exactly the same way as regression coefficients.* For example, the coefficient of price is approximately $-.95$, indicating that a one per cent increase in price leads, other things being equal, to a fall in spirits consumption of $-.95$. The 't-value' is -13.39. If the (relative) variances were known, this statistic would have a t-distribution. However, because the parameters are estimated all we can say is that it is normally distributed in large samples. The figure in square brackets gives the probability value for a two-sided test based on a standard normal distribution.

The basic results can thus be expressed in much the same way as in a classical regression:

$$y_t = \tilde{\mu}_{t|T} - \underset{(.071)}{.950}\, x_{1t} + \underset{(.132)}{.695}\, x_{2t} + \tilde{\varepsilon}_{t|T}, \quad t = 1, \ldots, T,$$

$$R_d^2 = 0.728, \quad \hat{\sigma} = 0.022, \quad Q(8,6) = 4.10, \quad N = 7.95.$$

where y_t is spirits, x_{1t} is price, x_{2t} is income and $\tilde{\mu}_{t|T}$ and $\tilde{\varepsilon}_{t|T}$ are the smoothed estimates of the trend and irregular components.

The above expression tells us nothing about the trend. To see precisely what the trend is doing, we move to **Test**/Components graphics. Mark 'Trend' and 'Trend plus Regression effects' in the 'Plots with Y' section. Press OK to get the graphs as displayed in Figure 5.3. The former shows the trend with the actual series. The gradual downward movement in the trend indicates a change in tastes away from spirits. Unlike a univariate series the trend does not pass through the observations. However, adding in the effect of the explanatory variables is shown in the second graph which reflects the equation as written above. (If we go back to the Components dialog and change 'Smoothed' to 'Filtered', this second graph shows one-step-ahead predictions. Although this is a standard output in econometric packages, we find it less useful than the plot of the prediction errors which is given in Residuals.)

As was noted earlier, the standard econometric approach to the SPIRIT data is to fit a deterministic trend and an AR(1) disturbance. To do this, enter the **Model**/Formulate dialog first and include the Y and X variables. In the Select components dialog, select 'Level, Fixed' and 'Slope, Fixed' (this is a fixed trend), select 'AR(1)' in the 'Cycle(s)' section and proceed as before. Compare the fit, coefficients and diagnostics with those obtained with the stochastic trend model.

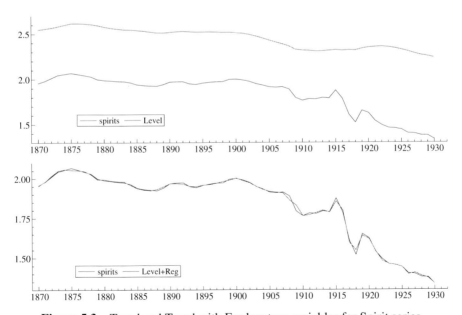

Figure 5.3 Trend and Trend with Explanatory variables for Spirit series.

5.2.2 Outliers and structural breaks

Intervention variables may appear in models together with explanatory variables. Similarly the auxiliary residuals may be used to detect the outliers and structural breaks which give rise to their inclusion. To illustrate, continue with the SPIRIT example. An examination of the output from Auxiliary residuals, indicates some outliers and a shift in the level in 1909; see Harvey and Koopman (1992) for further details.

In case STAMP is used with the option 'Select interventions automatically' activated in the Select components dialog, the following regression output is reported (based on the estimation sample 1870 – 1930):

```
Regression effects in final state at time 1930
```

	Coefficient	RMSE	t-value	Prob
Outlier 1918. 1	-0.06011	0.01054	-5.70292	[0.00000]
Level break 1909. 1	-0.09518	0.01484	-6.41588	[0.00000]
price	-0.82434	0.05563	-14.81940	[0.00000]
income	0.59481	0.10382	5.72920	[0.00000]

In case the full estimation sample 1870 – 1938 is used, we obtain:

```
Regression effects in final state at time 1938
```

	Coefficient	RMSE	t-value	Prob
Outlier 1915. 1	0.04525	0.00829	5.45911	[0.00000]
Outlier 1918. 1	-0.06177	0.00815	-7.58320	[0.00000]
Level break 1909. 1	-0.09499	0.01175	-8.08297	[0.00000]
price	-0.73800	0.04703	-15.69331	[0.00000]
income	0.67555	0.08072	8.36942	[0.00000]

These are also the intervention variables considered in Harvey and Koopman (1992) where it is discussed that the outliers are clearly associated with World War 1 and the level break is associated with the introduction of a law on spirits by prime minister Lloyd George.

5.2.3 Lags and differences

The SPIRIT example involved only current values of the explanatory variables. We can easily test whether lagged price should enter into the model. When the Formulate dialog is active, select 'price' and select 'lag' in the drop box of the 'Lags' section and type 1 in the box below. Press << and both 'price' and price lagged one period — 'price_1' — now appear as explanatory variables. Estimate the model with the same Components specification as before and carry out a 't-test' on lagged price.

The above example created lags within the Formulate dialog. However, lags can also be created in the Calculator of OxMetrics. This is useful if we wish to retain the lagged variable in the data set for future use.

It is often desirable to reformulate a lag structure in terms of differenced observations. This is for both theoretical reasons and practical reasons because differenced observations typically lead to greater stability in estimation since they suffer less from multicollinearity. The differencing must be carried out before entering the Formulate dialog. The Calculator of OxMetrics may be used for this purpose. The Calculator also offers other types of time series transformations.

5.3 Forecasting

To make forecasts of future values of the dependent variable, we need corresponding values of the explanatory variables. The Forecasting dialog is given below.

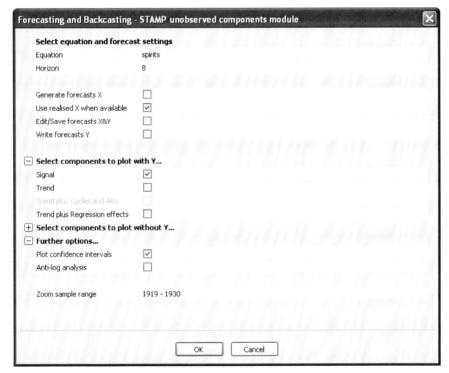

This dialog offers a flexible method of inputting future values for explanatory variables and changing them so that the path of the forecasts can be examined under different scenarios. We will continue with the SPIRIT example, and see how forecasts can be made from after the end of the estimation period in 1931 to 1938. Then within the **Test**/Forecasting dialog the 'Horizon' values remains at its default value of 8. Furthermore, select the options 'Signal' and 'Trend plus Regression effects'. The default zoom period of 1919 – 1930 can be changed but we keep the default. Pressing OK gives the default forecasts in which future values of the explanatory variables (1931 – 1938) are simply set to their values inside the database; see the upper graph of Figure 5.4. Some more interesting possibilities are outlined below.

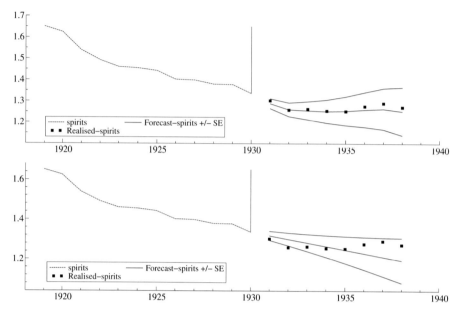

Figure 5.4 Forecasts with Explanatory variables for Spirit series: in upper graph, future Xs are replaced by their realised values in the database; in lower graph, future Xs are extrapolated from their values in period 1930 with an increment of 0.01 for each period.

5.3.1 Incremental change

In the Forecasting dialog, the 'Forecast settings' section has the options 'Generate forecasts X' and 'Edit/Save forecasts X&Y' to provide means of manipulating the values of the explanatory variables each period beyond the database sample. When the option 'Generate forecasts X' is activated, the following dialog appears that allows you to increase explanatory variables by a constant value each period in the forecasting horizon.

5.3 Forecasting

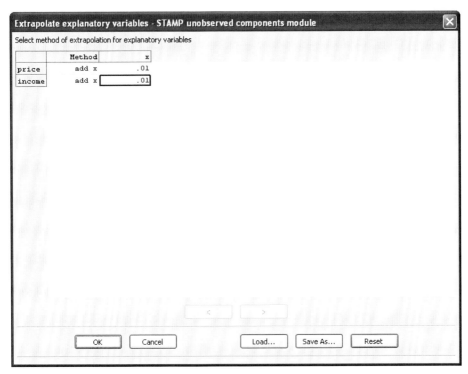

If the data are in logarithms, the increment is (one hundredth of) the growth rate per period. Thus, if in the present example, we feel that income and price should grow at 1% per year, '0.01' should be added each time period. By changing the values in the 'x' column to 0.01, both explanatory variables are increased by 0.01 in the forecasting period starting with their corresponding values in 1930. By pressing OK, the graphs of the series together with their forecasts appear. This dialog also allows one to generate future values for explanatory variables by other methods. The column 'Method' has drop boxes to choose methods of generating values for the forecasting period including percentage increases from the last observed values, means of the last x values of the explanatory variables and fixed trends based on the last x values. After generating future values for explanatory variables using this dialog, pressing OK leads to graphs of the series together with their forecasts. The graphs of the forecasts of the explanatory variables allow the user to judge whether the assumed extrapolations of the explanatory variables are plausible; see Figure 5.4.

5.3.2 Manual input

Future values of the explanatory variables can also be entered manually and this can be achieved as follows. Go to the Forecasting dialog and activate the option 'Edit/Save forecasts X&Y' in the 'Forecast settings' section. After selecting other options in the

Forecasting dialog and pressing OK, the Matrix Editor dialog appears that allows the user to change the forecast values of the explanatory variables in standard ways. The Matrix Editor has standard spreadsheet functionalities but it also allows saving the matrix of values on disk and loading values from files saved earlier on disk. In our case, you can edit, save or load the eight forecasts for price and for income. When finishing editing, the button OK can be pressed and the corresponding forecasts of spirits are presented in OxMetrics.

5.3.3 Using models to forecast the explanatory variables

Future values of the explanatory variables may be constructed by fitting models to them. Having fitted time series models, such as local linear trends, to each "explanatory" variable, forecasts can be made using the Forecasting dialog and then saved when the option 'Write forecasts Y' is activated. Suppose in the SPIRIT example that the forecasts of price and income are outputted in the Results window using the option 'Write forecasts Y' in the Forecasting dialog. Once the model has been estimated for spirits, go to Forecasting, select option 'Edit/Save forecasts X&Y' and you can reload the model-based forecasts of the explanatory variables using standard copy/paste facilities.

5.3.4 Interventions

If intervention variables are constructed using the Intervention dialog they will be automatically extended into the forecast period. Interventions constructed in any other way, such as by using the Calculator, must be treated in the same way as any other explanatory variable.

5.4 Statistical features of the models

Some insight into statistical aspects of the models can be obtained by examining a simple model with a single explanatory variable, a random walk trend (level), and irregular and level interventions. The model may be written

$$\begin{aligned} y_t &= \mu_t + \delta x_t + \lambda w_t + \epsilon_t, & \epsilon_t &\sim \text{NID}(0, \sigma_\epsilon^2), & t = 1, \ldots, T, \\ \mu_t &= \mu_{t-1} + \theta z_t + \eta_t, & \eta_t &\sim \text{NID}(0, \sigma_\eta^2), \end{aligned}$$

where w_t takes the value one at the time an event occurs and is zero otherwise. The specification of z_t is similar.

As noted in the introduction to the chapter, an alternative way of capturing the change in level induced by z_t is to put a step variable in the measurement equation, that is add another w_t variable which is zero before the event in question and unity thereafter. The attraction of the specification written above is that the step change is built into the trend component; see the Nile example in §5.1.

Now consider the model without the interventions and observe the following:

- If σ_η^2 is zero the level is a constant and the model reduces to a classical regression.
- If σ_ε^2 is zero, but σ_η^2 is not, the model is a standard regression in first differences, that is
$$\Delta y_t = \delta \Delta x_t + \eta_t, \quad t = 2, \ldots, T.$$
- If both variances are non-zero, the disturbance in the first difference equation is a first-order MA. In levels the stochastic part of the model is ARIMA(0,1,1). Hence the 'reduced form' of the stochastic trend regression model is what is sometimes called a 'transfer function' model; see Harvey (1989, Ch. 7) or Harvey (1993, Ch. 5) for further discussion.

5.5 Exercises

(1) Estimate the Nile model with interventions as specified in §5.1.1. Carry out a post-intervention test on the 1899 level shift based on the observations up to and including 1905.
(2) The new Aswan dam on the Nile was constructed in 1955. Use the NILE data to see if there is any impact on the flow of the river.
(3) Use the auxiliary residuals to detect outliers in the series of US exports to Latin America, LAXQ. Fit an appropriate model — you might perhaps specify a smooth trend and include a cycle.
(4) Re-estimate 'ofuCOAL1' in ENERGY with interventions for the miners' strike in 84, Q3 and Q4. How are the forecasts affected?
(5) Test if 'law' is significant in SEATBELT for 'Front' and 'Rear' passengers.
(6) An example of a dynamic model is the employment-output equation fitted to quarterly, seasonally adjusted, UK data by Harvey, Henry, Peters and Wren-Lewis (1986). This is contained in the file EMPL. The dependent variable is 'Empl'. Include a stochastic trend and two lags on the dependent variable and the explanatory variable, which is manufacturing output, 'Output'. Test the significance of the lags and hence simplify the model.
(7) In the book by Fuller (1996, p. 522), a quadratic time trend is fitted to the SPIRIT data. Such a variable may be constructed as indicated by forming an index using Algebra, and then squaring it. Fit a quadratic trend, with and without an AR(1) disturbance, and compare the fit with the stochastic trend model fitted in §5.2.1.

Chapter 6
Tutorial on Multivariate Models

6.1 SUTSE models

Seemingly unrelated time series equations (SUTSE) have a similar form to univariate models, except that $\mathbf{y_t}$ is now an $N \times 1$ vector of observations, which depends on unobserved components which are also vectors. The link across the different series is through the correlations of the disturbances driving the components. In a common factor model, some, or all, of the variance matrices will be of reduced rank. Common factors are of considerable importance, and their interpretation and how they may be imposed is discussed in detail in §6.4.

The fitting of multivariate models proceeds much as in the univariate case. The Formulate dialog in the Model menu (Alt+y) allows the marking of multiple variables as dependent (Y). This is done by selecting the appropriate variables in the Database listbox: use the mouse and double-click on a variable or highlight a variable and use the $<<$ keys to transfer the variables to the Selection listbox. Once the variables have been transferred to the Selection listbox, the first variable is marked as Y. It suggests that the default model is univariate and the other selected variables are regarded as explanatory variables. By right-clicking the mouse on a variable in the Selection listbox, the status of this variable can be changed to Y variable. This can also be achieved by highlighting the appropriate variables (for multiple highlighting, Ctrl and left-clicking the variables) in the Selection listbox, choosing Y variable in the droplist below the Selection listbox and pressing the button 'Set'. By pressing OK, the Select components dialog is entered. The default model is the multivariate basic structural time series model with full variance matrices for the disturbance vectors driving the unobserved component vectors. These default settings can be confirmed by pressing OK and the model is estimated in the usual way.

The output for each equation is listed in turn in the Results window. The output that appears in univariate modelling is presented for each variable in the dependent vector of time series. Furthermore, for each unobserved component the 'variance/correlation matrix' is presented as default output in the Results window: the diagonal elements are the variances, the lower triangular elements are the covariances and the upper triangular elements are the corresponding correlations. Plots of joint components can easily be produced within OxMetrics. The default component graphs can be found in the Model graphics window (in the Document list of OxMetrics, the left-hand side of the main program).

As an example, load the UKCYP data and build a multivariate BSM for consumption and income; having marked both 'c' and 'y' as dependent Y and then formulate the components model by following the defaults. The hyperparameter results are as follows. The elements of the estimated variance/correlation matrices for the irregular and level noise are:

$$\widehat{\Sigma}_\varepsilon = \begin{pmatrix} 6.11 \times 10^{-6} & \textit{-0.835} \\ -1.15 \times 10^{-5} & 3.11 \times 10^{-5} \end{pmatrix}, \widehat{\Sigma}_\eta = \begin{pmatrix} 4.98 \times 10^{-5} & \textit{0.933} \\ 8.55 \times 10^{-5} & 1.69 \times 10^{-4} \end{pmatrix}.$$

The elements in italics are the estimated correlations. The correlation is much higher for the disturbances in the levels than for the irregular. The corresponding matrices for the slope and seasonal are:

$$\widehat{\Sigma}_\zeta = \begin{pmatrix} 2.84 \times 10^{-6} & \textit{1} \\ 2.05 \times 10^{-6} & 1.48 \times 10^{-6} \end{pmatrix}, \widehat{\Sigma}_\omega = \begin{pmatrix} 1.28 \times 10^{-6} & \textit{-0.213} \\ -4.37 \times 10^{-7} & 3.28 \times 10^{-6} \end{pmatrix}.$$

The slope disturbances are perfectly correlated while the correlation between the seasonals is negative and close to zero.

In the special case of a multivariate local level model

$$\begin{aligned} y_t &= \mu_t + \epsilon_t, & \epsilon_t &\sim \text{NID}(0, \Sigma_\epsilon), \\ \mu_t &= \mu_{t-1} + \eta_t, & \eta_t &\sim \text{NID}(0, \Sigma_\eta), \end{aligned} \quad (6.1)$$

where Σ_ϵ and Σ_η are the $N \times N$ variance matrices, and η_t and ϵ_t are multivariate normal disturbances which are mutually uncorrelated in all time periods. For the more general multivariate BSM, the other disturbances similarly become vectors which have $N \times N$ variance matrices. In the case of trigonometric seasonals there are two sets of $N \times 1$ vectors for each seasonal frequency such that

$$E(\omega_{it}\omega'_{it}) = E(\omega^*_{it}\omega^{*'}_{it}) = \Sigma_\omega, \quad E(\omega_{it}\omega^{*'}_{it}) = 0, \quad i = 1, 2, \ldots, [s/2],$$

and all disturbances at different frequencies are independent.

In STAMP 8 we allow for a range of variance matrices structures that can be selected for each unobserved component. The following variance matrices can be selected:

- *full*: a full variance matrix of rank N is considered as in the UKCYP example above;
- *scalar*: variance matrix Σ is specified as the unity matrix scaled by a non-negative value, that is $\Sigma = \sigma^2 I_N$ where $\sigma^2 \geq 0$ is a scalar variance;
- *diagonal*: a diagonal variance matrix is considered with N different diagonal elements;

- *ones*: variance matrix Σ is specified as a matrix of ones scaled by a non-negative value, that is $\Sigma = \sigma^2 \iota\iota'$ where $\sigma^2 \geq 0$ is a scalar variance and ι is a vector of ones (note: Σ has rank one in case $\sigma^2 > 0$);
- *cdiag* : variance matrix Σ is specified as a $\Sigma = aa' + D$ where a is an $N \times 1$ non-zero vector and D is a diagonal matrix.

6.2 Cycles

Cycles may be introduced into multivariate models. As with other components, the disturbances can be correlated across the series. Since the cycle in each series is driven by two disturbances, there are two sets of disturbances and these are assumed to have the same variance matrix, that is

$$E(\kappa_t \kappa_t') = \mathbf{E}(\kappa_t^* \kappa_t^{*'}) = \Sigma_\kappa, \quad \mathbf{E}(\kappa_t \kappa_t^{*'}) = \mathbf{0}, \quad t = 1, \ldots, T,$$

where Σ_κ is an $N \times N$ variance matrix. The homogeneity restriction can be applied to models with cycles if desired.

As in the univariate case, STAMP allows up to three cycles in each series, but it imposes the restriction that, for a given cycle, *the damping factor and the frequency, ρ and λ_c, are the same for all series*. This means that the cycles in different series have the same properties, that is they have the same autocorrelation function and spectrum. We call them *similar* cycles. The strength of a cycle in a particular series depends on the variance of its disturbance.

The file MINKMUSK contains two series showing the numbers of furs of minks and muskrats traded annually by the Hudson Bay Company in Canada from 1848 to 1909. A model may be fitted with similar cycles in order to establish the joint stylised facts associated with the two series. Proceed by selecting the logarithms of the two series (Lmink and Lmuskrat) and select 'Fixed' for Level component and select 'Cycle medium (default 10 years)' (keep the other options at their default settings) in the Select components dialog. The last three observations are best omitted as they are somewhat atypical. Thus the end of the sample in the Estimate dialog should be changed to 1906.

Some relevant output for the cycle is given by

```
Variances of disturbances in Eq Lmink:
                   Value    (q-ratio)
Level           0.000000   (   0.0000)
Slope           5.11905e-005 ( 0.001427)
Cycle           0.00137809 (   0.03842)
Irregular       0.0358708  (   1.000)
Variances of disturbances in Eq Lmuskrat:
                   Value    (q-ratio)
Level           0.000000   (   0.0000)
```

```
Slope             0.000251142  ( 0.007001)
Cycle             0.113689     (   3.169)
Irregular         0.0281232    (  0.7840)

Cycle other parameters:
Period              9.78101
Frequency           0.64239
Damping factor      0.98448
Order               1.00000
Cycle variance/correlation matrix:
              Lmink      Lmuskrat
Lmink         0.04475    -0.3049
Lmuskrat     -0.02283     0.1253
```

The estimated trends are relatively smooth, not unlike the quadratics fitted by Chan and Wallis (1978) in their initial detrending procedure. The smoothness arises because the level variances were constrained to be zero and the q-ratios for the slope are small. The cycles have a period parameter of 9.78 years. The relative importance of the cycles is indicated by State vector analysis. For data in logarithms, the amplitude of each cycle is a percentage of the trend. This is for Lmink 16% and for Lmuskrat 34%.

```
State vector analysis at period 1906
Equation Lmink
                         Value       Prob
Level              10.89963 [0.00000]
Slope               0.00116 [0.95299]
Cycle 2 amplitude   0.16325 [     -]

Equation Lmuskrat
                         Value       Prob
Level              13.80520 [0.00000]
Slope               0.02754 [0.51726]
Cycle 2 amplitude   0.34475 [     -]
```

When the two estimated cycles are stored into the database of MINKMUSK, OxMetrics provides graphical tools to plot the two cycles together; given in Figure 6.1. In the dialog **Test**/Components graphics, the boxes for 'Trend' and 'Irregular' should be de-activated and the box for 'Cycles and ARs' should be activated. By selecting the option 'Store selected components in database' in 'Further options...', the program will prompt you to give a name for the series to be stored. It is suggested to keep the default names. Using the **Model**/Graphics option in OxMetrics, the two series can be graphed jointly and also a cross-plot can be drawn.

Using the **Model**/Calculator option in OxMetrics, the series can be shifted backwards and forwards; this is done by taking lags of the series. After some experimenta-

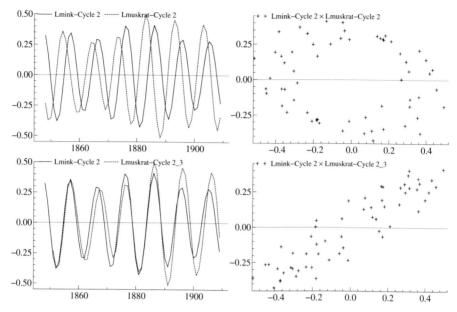

Figure 6.1 The estimated cycles for minks and muskrats.

tion it will be found that the Lmuskrat cycle leads the mink cycle by about three periods. (Take the Lmuskrat series with lag 3). The correlation between the cycle Lmink and the cycle Lmuskrat with lag 3, as shown in Figure 6.1, is 0.9. These results are consistent with the findings of Chan and Wallis.

It is interesting that the variance matrix of the slope variance is estimated to be singular; the meaning of this will be explored in §6.4. You may want to save the joint plots of 'Trend' and 'Slope' using the Graphics dialog of OxMetrics.

The predictions within the sample are very good, but there is predictive failure for the three observations outside the sample — for some reason these appear very different. To make forecasts of future observations using the last three observations, re-estimate the model using the Estimate dialog.

6.3 Autoregression

A stationary first-order VAR may be included in a model as an alternative to, or even as well as, a cycle. The possible lag orders are 1 and 2. The coefficient matrices of the lag polynomials are assumed diagonal while the variance matrix of the disturbance vector can have any specification as given at the end of Section 6.1.

An example of a VAR(2) with a particularly useful interpretation is obtained by a minor modification of the mink and muskrat model fitted in the previous section. In the Select components dialog select 'AR(2)' instead of 'Cycle medium (default 10 years)'.

Thus
$$y_t = \mu_t + \psi_t + \epsilon_t, \quad \text{where,}$$
$$\psi_t = \Phi_1 \psi_{t-1} + \Phi_2 \psi_{t-2} + \kappa_t, \quad (6.2)$$

where Φ_1 and Φ_2 are diagonal matrices. After estimating this model, the Results window gives the coefficients of the VAR; that is, the elements of Φ_1 and Φ_2. These coefficients are:

$$\widehat{\Phi_1} = \begin{pmatrix} 0.754 & 0 \\ 0 & 1.540 \end{pmatrix}, \quad \widehat{\Phi_2} = \begin{pmatrix} -0.142 & 0 \\ 0 & -0.593 \end{pmatrix}.$$

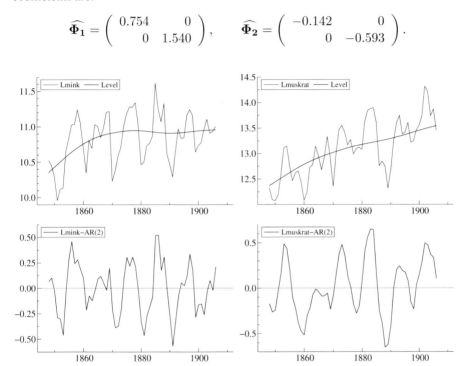

Figure 6.2 The estimated trend and autoregressive components for minks and muskrats.

The VAR(2) component is a multivariate time series process that is able to generate cycles of the form observed earlier; see Figure 6.2. The forecasts show the cycles continuing into the future, but damping down.

6.4 Common factors and cointegration

In a common factor model, some or all of the components are driven by disturbance vectors with less than N elements. *Recognition of common factors yields models which may not only have an interesting interpretation, but may also provide more efficient inferences and forecasts.* Many of the applications described in the next chapter involve common factors and their interpretation.

In terms of a SUTSE model, the presence of common factors means that the variance matrices of the relevant disturbances are less than full rank. The common factor restrictions can be imposed in the Select components dialog by selecting 'Multivariate settings...'. In the Select variance matrices and components for each equation dialog, appropriate variance matrices can be selected and their ranks can be determined.

The first subsection below describes the statistical specification of the common levels model. Common slopes, seasonals and cycles may be formulated along similar lines. Full details can be found in Chapter 9. The irregular variance matrix may also be less than full rank, though this does not, as a rule have a particularly useful interpretation, and there is rarely a case for specifying it as such on prior grounds.

6.4.1 Statistical specification of common levels

Consider the local level model, (6.1), but suppose that the rank of Σ_η is $K < N$. The model then contains K common levels or *common trends* and may be written as

$$\begin{aligned} y_t &= \Theta \mu_t^\dagger + \mu_\theta + \epsilon_t, & \epsilon_t &\sim \text{NID}(0, \Sigma_\epsilon), \\ \mu_t^\dagger &= \mu_{t-1}^\dagger + \eta_t^\dagger, & \eta_t^\dagger &\sim \text{NID}(0, \Sigma_\eta^\dagger), \end{aligned} \quad (6.3)$$

where $\eta_t^\dagger$ is a $K \times 1$ vector, Θ is an $N \times K$ matrix of *(correlated) factor loadings*, $\Sigma_\eta^\dagger$ is a variance matrix of full rank K and μ_θ is an $N \times 1$ constant vector. The Θ matrix consists of K rows that span the identity matrix I_K. The remaining $N - K$ rows of Θ are contained in the matrix $\overline{\Theta}$. The constant vector μ_θ has K zero values; the $N - K$ non-zero elements are contained in a vector $\overline{\mu}$. A typical example is

$$\Theta = \begin{bmatrix} I_K \\ \overline{\Theta} \end{bmatrix}, \quad \mu_\theta = \begin{pmatrix} 0 \\ \overline{\mu} \end{pmatrix}. \quad (6.4)$$

The program allows a re-ordering of the rows of Θ: if a series is set to 'dependent' in the Select variance matrices ... dialog, it is assigned to the last $N - K$ rows. The same re-ordering then applies to $\overline{\mu}$. The next sub-section provides an illustration,

The model may be recast in the original SUTSE form (6.1) by writing $\mu_t = \Theta \mu_t^\dagger + \mu_\theta$ and noting that $\Sigma_\eta = \Theta \Sigma_\eta^\dagger \Theta'$ is a singular matrix of rank K. In case of (6.4), we have

$$\Sigma_\eta = \Theta \Sigma_\eta^\dagger \Theta' = \begin{bmatrix} \Sigma_\eta^\dagger & \Sigma_\eta^\dagger \overline{\Theta}' \\ \overline{\Theta} \Sigma_\eta^\dagger & \overline{\Theta} \Sigma_\eta^\dagger \overline{\Theta}' \end{bmatrix}, \quad \mu_t = \Theta \mu_t^\dagger + \mu_\theta = \begin{pmatrix} \mu_t^\dagger \\ \overline{\mu} + \overline{\Theta} \mu_t^\dagger \end{pmatrix}.$$

The $K \times K$ variance matrix $\Sigma_\eta^\dagger$ is decomposed by the Cholesky decomposition

$$\Sigma_\eta^\dagger = \mathbf{LDL}', \quad (6.5)$$

where $\mathbf{L}$ is a $K \times K$ unity lower triangular matrix and $\mathbf{D}$ is a $K \times K$ diagonal positive matrix. The lower coefficients in $\mathbf{L}$ and the diagonal elements of $\mathbf{D}$ are estimated together with elements in $\overline{\Theta}$ for each variance matrix associated with a disturbance vector. When there are no common trends, so $N = K$, the coefficient matrix $\overline{\Theta}$ is a null matrix and matrices **L** and **D** are for the Cholesky decomposition of Σ_η. Also, when $N = K$ vector $\overline{\mu}$ is null and $\mu_t = \mu_t^\dagger$.

The presence of common trends implies what econometricians call *cointegration*. While the observations are integrated of order one; that is, they must be differenced once to make them stationary, there exists an $(N - K) \times N$ matrix of cointegrating vectors, **A**, such that $\mathbf{A}\mathbf{y}_t$ is stationary. This means that $\mathbf{A}\Theta = \mathbf{0}$; see the discussion in Harvey (1989, Ch. 8). Testing procedures for common trends in multivariate structural time series models have been developed by Nyblom and Harvey (2001).

6.4.2 An example of common levels: road casualties in Britain

The file SEATBQ contains quarterly time series on drivers and rear seat passengers killed and seriously injured in road accidents in Great Britain. These are labelled 'Drivers' and 'Rear' in SEATBQ. Select them both in the Formulate dialog and label them both as 'Y' (right-click Rear to set it as a 'Y' variable, the same as Drivers). A graph of the series indicates that the local level is appropriate for the trend, so deselect 'Slope' in the Select components dialog (click the 'Slope' checkbox so the tick disappears). A seasonal must also be included, but we shall ignore it in the discussion. Estimate the model with no restrictions, but for reasons which will become apparent in §6.6, only use the observations up to the end of 1982. Thus, set 'Estimation ends at' to $82(4)$ in the Estimate dialog. The parameter estimates of the covariance/correlation matrix for the level disturbance is estimated by

```
Level disturbance variance/correlation matrix:
             Drivers         Rear
Drivers      0.0005772       0.8908
Rear         0.0003919       0.0003353
```

The high correlation suggests that there may be a common factor. The 'Variances' option in the dialog **Test**/**More** written output also reveals that

```
Cholesky decomposition LDL' with L and D given by
             Drivers         Rear
Drivers        1.000         0.0000
Rear           0.6790        1.000
diag(D)      0.0005772       6.922e-005
Eigenvectors and eigenvalues are given by
             Drivers         Rear
Drivers        0.8046        0.5938
Rear           0.5938       -0.8046
eigenvalues  0.0008664       4.611e-005
```

The small eigenvalue for Rear suggests that the rank of the Level disturbance variance matrix may be one.

Now return to the Select components dialog and ensure that the option 'Multivariate settings' is selected. In the dialog Select variance matrices and components for each equation, the level component for Rear can be made fully dependent on the trend for Front by changing the entry 'in' of level (row) and Rear (column) to 'dependent'. This can be achieved by double-clicking on this entry and selecting the appropriate label 'dependent'. In effect, the rank of the level disturbance variance matrix has changed from '2' to '1'.

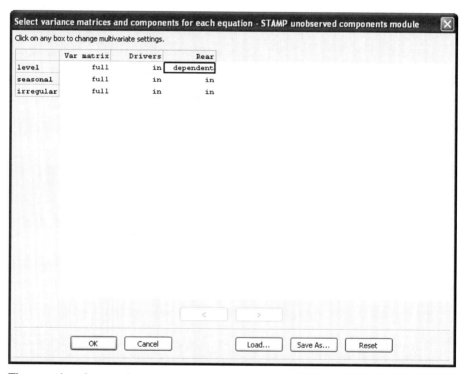

Then continue by pressing OK. The covariance matrix of the level disturbances is now given by

```
Level disturbance variance/correlation matrix:
            Drivers          Rear
Drivers   0.0005690         1.000
Rear      0.0004491         0.0003544
Level disturbance factor variance for Drivers: 0.000568957
Level disturbance factor loading  for Rear: 0.789258
            Drivers          Rear
Constant    0.0000           0.08107
```

Ignoring the seasonals, the estimated model may be written as:

$$y_{1t} = \mu_t^\dagger + \varepsilon_{1t},$$
$$y_{2t} = 0.789\mu_t^\dagger + 0.0811 + \varepsilon_{2t},$$

where $\mu_t^\dagger$ is a *univariate* random walk. Thus we have the following relationship between the level components as they appear in the two series:

$$\mu_{2t} = 0.789\mu_{1t} + 0.0811.$$

Since the data are in logs, the relationship between the original trends is:

$$Trend_2 = \exp(0.0811) Trend_1^{0.789}.$$

Finally note that the system is cointegrated of order $(1, 1)$, with cointegrating vector $(1, -1/.789) = (1, -1.267)$.

6.4.3 Balanced levels

In a local level model with a single common factor, the trends in the series will only be parallel (or proportional if the data are modelled in logs and anti-logs have been taken) if the elements of the standardised load matrix, Θ, are all unity. Since the first element is always unity, this implies $N-1$ restrictions. These *balanced level* restrictions may be imposed in two ways. The easiest way is go back to the Select components dialog and ensure that the option 'Multivariate settings' is selected. In the dialog Select variance matrices and components for each equation, reset the level entry under 'Rear' to 'in' and set the 'Var matrix' to 'ones' (double-click on the entry 'full' for level 'Var matrix' and select 'ones'). The estimated level disturbance variance matrix is given by

```
Level disturbance variance matrix of ones, scaled:
            Drivers         Rear
Drivers     0.0005011       0.0005011
Rear        0.0005011       0.0005011
```

The output in the Results window indicates that the fit is excellent.

```
Summary statistics
                Drivers         Rear
T               56.000          56.000
p               2.0000          2.0000
std.error       0.082540        0.10451
Normality       3.9943          1.2563
H(17)           0.70313         0.74554
DW              2.0218          2.1373
r(1)            -0.023856       -0.087228
q               8.0000          8.0000
```

```
r(q)                -0.10296    -0.14327
Q(q,q-p)             9.2697      2.6926
Rs^2                 0.49474     0.48793
```

In this case, an alternative way to enforce the restriction of a factor loading of one for the level of Rear is to keep its level as dependent and enforce the unity factor loading via the option 'Set parameters to default values and edit' in the Select components dialog (click on the appropriate radio button which can be found in the dialog below). In the dialog Select variance matrices and components for each equation we select a full 'Var matrix' for level and a 'dependent' level for Rear, as before. In the next dialog Edit and fix parameter values, click on the 'Fix' checkbox of 'Factor loading Level' and change its corresponding 'Value' from 0 to 1.

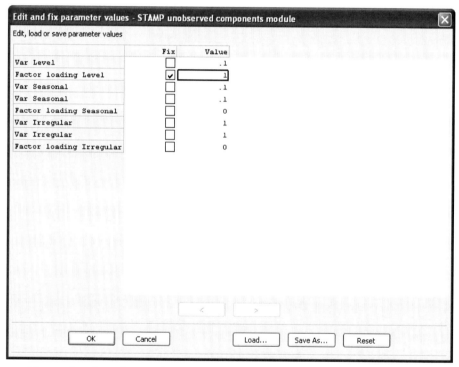

After estimation, the results are effectively the same as it should be. The level disturbance variance matrix is reported slightly differently in this case

```
Level disturbance variance/correlation matrix:
               Drivers         Rear
Drivers     0.0005011        1.000
Rear        0.0005011        0.0005011
Level disturbance factor variance for Drivers: 0.000501133
Level disturbance factor loading  for Rear: 1
```

```
                Drivers        Rear
Constant        0.0000         -1.491
```

The factor loading matrix Θ for the level is now just a column of ones while the estimate of $\bar{\mu}$ is -1.491. A plot of both trends shows them to be parallel. In the unrestricted model estimated in the previous subsection this was not the case. However, in both models the forecast functions are parallel. In the unrestricted model, the estimate of the second element in the factor loading matrix was 0.789. The likelihood ratio statistic for the hypothesis that the true value is one is $LR = 246.089 - 245.662 = 0.427$. Thus the hypothesis of a unit factor loading is easily accepted.

6.4.4 Common trends

Common levels is a special case of common trends as it arises when the trend is just a multivariate random walk. Models with common slopes may be formulated along similar lines. The full implications of different combinations of ranks of level and slope disturbances are laid out in Chapter 9. Here we just look at an important special case: *smooth trends with common slopes*.

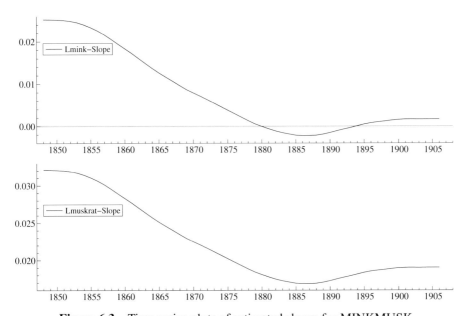

Figure 6.3 Time series plots of estimated slopes for MINKMUSK.

The simplest case to appreciate is where the variance matrix of the slope disturbances is less than full rank but the variance matrix of levels is null so that the estimated trends are relatively smooth. The mink–muskrat example of section 6.2 showed such a situation. Recompute the model (trend plus medium cycle plus irregular) imposing the constraint that the level is fixed and the slope of Lmuskrat is 'dependent' so that the rank of the slope disturbance variance matrix is one. Do not forget to change the estimation sample by typing '1906' in the entry for 'Estimation ends at' in the Estimate dialog. The output shows the standardised load matrix is $(1, \theta)' = (1, 0.555)'$ while the estimate of $\overline{\beta}$ is 0.018. Thus the growth rate in muskrats; that is, the slope in the second series, is given by :

$$\beta_{2t} = \theta \beta_{1t} + \overline{\beta} = 0.555 \beta_{1t} + 0.018.$$

The graph of the slopes shows this exact linear relationship; see Figure 6.3. Note that the muskrat growth rate is higher even though its stochastic growth component is less. The fitted model may be written:

$$\begin{aligned} y_{1t} &= \mu_t^\dagger + \varepsilon_{1t}, \\ y_{2t} &= 0.555 \mu_t^\dagger + 7.6 + 0.018t + \varepsilon_{2t}, \end{aligned}$$

where the constant term in the second (muskrat) equation is calculated from the levels given in the Final state; that is, since

$$\overline{\mu} = -\theta \mu_{1T} + \mu_{2T},$$

its estimate is $-0.555(10.93) + 13.67 = 7.6$.

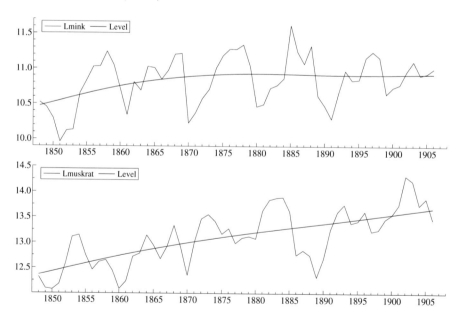

Figure 6.4 Time series plot of estimated trends for MINKMUSK.

As regards the relationship between the two trends:

$$\mu_{2t} = \theta \mu_{1t} + \overline{\mu} + \overline{\beta}t = 0.555 \mu_{1t} + 7.6 + 0.018t,$$

and so one trend is a multiple of the other plus a deterministic trend component; see Figure 6.3. It is only the stochastic movements in the trends which are in common. Finally note that the system is cointegrated of order $(2, 2)$; that is, the series are integrated of order 2 and there is a combination of them which is stationary. The cointegrated trends are displayed in Figure 6.4.

Although not appropriate here, it should be noted that balanced slopes ($\theta = 1$) can be imposed by setting the slope 'Var matrix' to 'ones'.

6.4.5 Common seasonals

Common factors in seasonality imply a reduction in the number of disturbances driving changes in the seasonal patterns. A single common factor does not imply that the seasonal patterns will be proportional unless the deterministic seasonal components outside the common seasonals are all zero. What it does mean is that the changes in the seasonal patterns in the different series come from a common source.

For trigonometric seasonals it is possible, in principle, to have different disturbance variance matrices for each of the seasonal frequencies, thereby allowing common factors in some frequencies but not in others. This implies seasonal cointegration at different frequencies; see Hylleberg, Engle, Granger and Yoo (1990). The current version of STAMP only allows full seasonal cointegration ; that is, common factors at all frequencies. Testing procedures for common seasonal factors are developed by Busetti (2006).

6.4.6 Common cycles

With *common cycles*, a vector of constant terms, corresponding to the non-zero elements in μ_θ for the level in (6.3), is not included since the expectation of a cycle is zero.

It should be clear that common cycles embody much stronger restrictions than similar cycles. Thus with two series and one common cycle, one cycle would be proportional to the other. Common cycles are like the common feature cycles of Engle and Kozicki (1993).

A test of the null hypothesis that there are a fixed number of common cycles can be based on the LR statistic obtained from fitting the model with the common cycle restriction and the (unrestricted) similar cycle model. The distribution theory is complicated by boundary conditions, but is described in Carvalho, Harvey and Trimbur (2007). When there are only two series, the appropriate significance point for a test at the 5% level of significance is given by the 10% critical value of a chi- square with one degree of freedom, that is 2.71.

6.4.7 Factor rotations

When there is more than one common factor, they are not unique and issues of interpretation arise. This leads on to factor rotations. Consider the local level model. The first step is to formulate the multivariate local level model (6.3) in such a way that the common factors are uncorrelated with each other and have unit variance, that is

$$y_t = \Theta^* \mu_t^* + \mu_\theta + \epsilon_t, \quad \epsilon_t \sim \text{NID}(0, \Sigma_\epsilon),$$

$$\mu_t^* = \mu_{t-1}^* + \eta_t^*, \quad \eta_t^* \sim \text{NID}(0, I_K),$$

where Θ^* is an $N \times K$ matrix of factor loadings given by $\Theta^* = \Theta L D^{1/2}$, the new components are $\mu_t^* = D^{-1/2} L^{-1} \mu_t^\dagger$ and L and D are from the Cholesky decomposition in (6.5). Since L is lower triangular, Θ is such that its elements, θ_{ij}, are zero for $j > i$ and $i = 1, ..., K$.

When there is more than one common factor, they are not unique and a *factor rotation* may give components with a more interesting interpretation. Let H be a $K \times K$ orthogonal matrix. The matrix of factor loadings and the vector of common trends can then be redefined as $\Theta^\ddagger = \Theta H'$ and $\mu_t^\ddagger = H \mu_t^\dagger$ yielding

$$\begin{aligned} y_t &= \Theta^\ddagger \mu_t^\ddagger + \mu_\theta + \epsilon_t, & \epsilon_t &\sim \text{NID}(0, \Sigma_\epsilon), \\ \mu_t^\ddagger &= \mu_{t-1}^\ddagger + \eta_t^\ddagger, & \eta_t^\ddagger &\sim \text{NID}(0, I_K). \end{aligned}$$

The disturbances driving the common trends are still mutually uncorrelated with unit variance.

A number of methods for carrying out rotations have been developed in the classical factor analysis literature. These may be employed here. The program does not offer an option for computing rotations at present, but for two factors a commonly used rotating matrix is

$$H = \begin{bmatrix} \cos \lambda & -\sin \lambda \\ \sin \lambda & \cos \lambda \end{bmatrix}$$

with the angle, λ, being set by a graphical method. The aim is often to give a factor significant loadings, perhaps all positive, on some variables while the other variables get loadings near zero; see Harvey, Ruiz and Shephard (1994) for a stochastic volatility application using the EXCH data.

6.5 Explanatory variables and interventions

Explanatory variables and interventions may be included in multivariate models. Thus

$$y_t = \mu_t + \gamma_t + \psi_t + \sum_{\tau=1}^{r} \Phi_\tau y_{t-\tau} + \sum_{\tau=0}^{s} D_\tau x_{t-\tau} + \Lambda w_t + \epsilon_t, \quad t = 1, ..., T,$$

where $\mathbf{x}_t$ is a $K \times 1$ vector of explanatory variables and $\mathbf{w}_t$ is a $K^* \times 1$ vector of interventions. Elements in the parameters matrices, $\mathbf{\Phi}, \mathbf{D}$, and $\mathbf{\Lambda}$ may be specified to be zero, thereby excluding certain variables from particular equations.

The next section gives an example of how explanatory variables and interventions can be included in a multivariate model.

6.6 Assessing the effect of the seat belt law using a control group

The effect of an intervention on a series can be modeled in a univariate framework as described in the previous chapter. However, suppose observations are available on a second series that is highly correlated with the series of interest but is not itself affected by the intervention. In this case it is possible to construct a bivariate model and so use the second series as a *control group*. This should result in a more precise measure of the intervention effect. The multivariate capability of STAMP offers the possibility of applying this technique.

The file SEATBQ was used earlier to construct a multivariate model of front and rear seat passengers killed and seriously injured (KSI) in road accidents in cars in Great Britain. Data on the number of kilometres travelled and the real price of petrol is also included in the file and Harvey and Durbin (1986) used these data in their study of the effect of the seat belt law of the first quarter of 1983.

Apart from the unavailability of suitable software, the main reason why Harvey and Durbin (1986) did not carry out intervention analysis with control groups was because of the lack of a suitable control group. Here we will use rear seat passengers as a control for front seat passengers. Although rear seat passengers were not required to wear belts by the law, it has been argued that they may not be a good control since:

(1) if, as the risk compensation theory implies, drivers wearing belts drive worse, there will be more rear seat passengers injured;
(2) rear seat passengers may sustain more serious injuries because those in the front are wearing belts and hence remain in their seats on impact;
(3) passengers might have transferred, or been transferred, to the rear so that a belt need not be worn;
(4) more rear seat passengers may have followed those in the front in wearing belts.

Despite these objections, there is little statistical evidence to suggest that the seat belt law had any effect on rear seat passengers KSI (try fitting an intervention model). Nevertheless, it should be borne in mind that the first three of the points above would all result in the effect of the law on the front seat occupants being exaggerated.

The procedure described in sub-section 6.4.3 is modified only slightly. In the Select components dialog, remove the slope component (deselect by clicking on checkbox

'Slope') as before and select 'manually...' under the option 'Select interventions'. Press OK. The Select interventions dialog appears. Set it so there is a level variable at 83(1). Then click on the box below 'Rear' and select 'out'. The dummy variable then only affects Drivers.

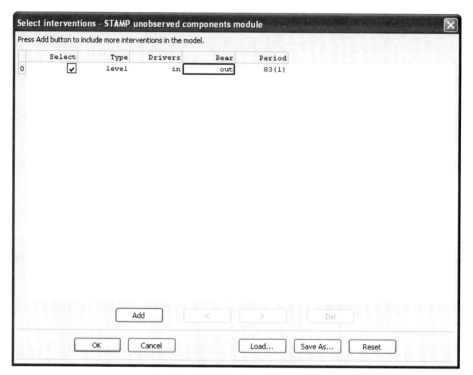

On estimating the model with the full sample the coefficient of the level dummy is found to be -0.240 with a t-statistic of -6.172. Figure 6.5 shows the two smoothed (balanced) levels with the intervention effect clearly showing for Drivers.

Explanatory variables, Kms and Petrol, may be added. Try this with 'Front' and 'Rear'. It will be found that Petrol is not statistically significant in the Rear equation, nor is Kms in the Front. They may be excluded by checking 'Set regression coefficients...' and changing the boxes under the appropriate variables to 'out'.

6.7 Exercises

(1) Construct a joint model for Drivers and Front seat passengers in SEATBQ. Try the common trend specifications. Extend the model to include the explanatory variables, Kms and Petrol.

(2) Estimate a bivariate model for MINKMUSK with a level included and the trend

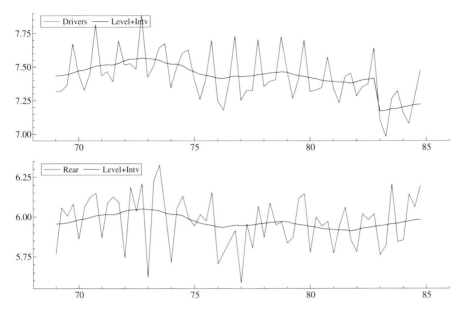

Figure 6.5 Smoothed levels of Drivers and Rear Seat Passengers killed and seriously injured in Great Britain, with allowance made for the seat belt law..

variance matrix restriction imposed of 'ones'. What is the implication of this restriction and how does the model compare with the one fitted in §6.2?

Chapter 7

Applications in Macroeconomics and Finance

This section gives some illustrations of the uses of STAMP. The focus will be on macroeconomics and finance, but the methods shown here could have easily been used in many other subjects.

7.1 Univariate trend-cycle decompositions: GDP

Model-based trend-cycle decompositions can be carried out by STAMP in different ways. The trend-cycle decomposition model is given by

$$y_t = \mu_t + \psi_t + \epsilon_t, \qquad \epsilon_t \sim \mathsf{NID}(0, \sigma_\epsilon^2), \qquad t = 1, \ldots, \mathsf{T},$$

where μ_t represents the trend, ψ_t the cycle and ϵ_t the irregular component. Different specifications of the three components are considered in the literature. The following typical specifications may be considered in STAMP and can be specified in the dialog **Formulate**/Select components:

- The trend is a random walk with fixed drift ('Level, Stochastic' and 'Slope, Fixed'), the cycle is an autoregressive component of order 2 ('AR(2)') and the irregular is white noise or zero; see, for example, Clark (1989).
- The decomposition model consists of a local linear trend ('Level, Stochastic' and 'Slope, Stochastic'), a stationary trigonometric cycle at a typical business-cycle frequency ('Cycle short' or 'Cycle medium') and an irregular; see Harvey (1989) for a more detailed exposition.
- The trend is imposed to be smooth by having 'Level, Fixed', 'Slope, Stochastic' and possibly with value of 'Order of trend' larger than 1 in the trend specification (4.6); the cycle is generalised as in (4.11) where 'Order of cycle' k is typically chosen as 2 or 3 and the irregular is the standard white noise sequence; see Harvey and Trimbur (2003).

Mixtures and extensions of these specifications can also be considered. For example, the cycle component can be the sum of two cycles at different frequencies.

7.1 Univariate trend-cycle decompositions: GDP

The parameters of the different trend-cycle decomposition models are estimated by maximum likelihood. The filtering properties are revealed by weight functions (time domain) and gain functions (frequency domain). To illustrate the facilities provided by STAMP, we consider the US GDP quarterly time series from 1947(1)-2007(2) (in logs) from the database file 'USmacro07.in7'. The three basic trend-cycle decomposition models can be estimated by maximum likelihood. However, in case the generalised cycle is included, the business-cycle frequency may need to be enforced. This is illustrated below.

The smooth trend plus generalised cycle plus irregular model is considered for logged US GDP below. Select the time series 'LGDP' from the database in the Formulate dialog, press OK and select 'Level, Fixed', 'Slope, Stochastic' and keep 'Order of trend' to its value 1. De-select 'Seasonal', keep 'Irregular' and expand the section 'Cycle(s)'. Select 'Cycle short' and increase the value of 'Order of cycle to 3. In section 'Options', select 'Edit and fix parameter values' and press OK. The Edit and fix parameter values dialog appears:

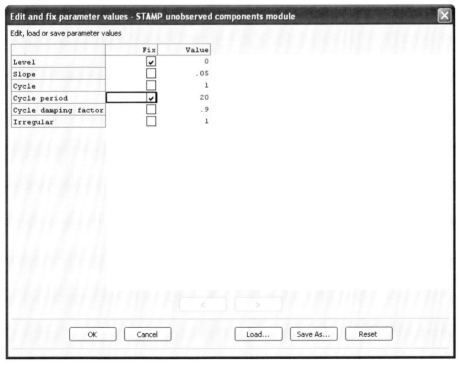

It presents the parameters that need to be estimated and their associating (starting) values. The first column has the name of the parameter, when only the name of the component appears, the parameter is the variance of this component. The value of the variances are relative to each other. In other words, the q-ratios appear where the q-ratios for the trend-cycle model are given by

$$q_\eta = \sigma_\eta^2 / \sigma^2, \quad q_\zeta = \sigma_\zeta^2 / \sigma^2, \quad q_\kappa = \sigma_\kappa^2 / \sigma^2, \quad q_\epsilon = \sigma_\epsilon^2 / \sigma^2,$$

where σ^2 is any of the variances σ_η^2, σ_ζ^2, σ_κ^2 or σ_ϵ^2, the one with the largest value is typically chosen.

By default, the largest variance is set to one and is typically σ_ϵ^2. The Edit and fix parameter values dialog also allows the user to fix parameter values. Since the dialog Select components has the level equation set by 'Level, Fixed', the variance of the 'Level' is fixed at zero here. In the Edit and fix parameter values dialog you can effectively change the level equation by changing its variance value to a nonzero value and re-activate it for estimation.

The estimation of this trend-cycle model usually requires fixing the cycle frequency to a typical business-cycle frequency of five years or twenty quarters. Therefore, the cycle frequency is fixed at 20 in this dialog by clicking on the box in the column 'Fix' and the row 'Cycle period'. Press OK to start the estimation process by pressing OK in the Estimation dialog. We obtain the following output:

```
UC( 1) Estimation done by Maximum Likelihood (exact score)
      The database used is USmacro07.in7
      The selection sample is: 1947(1) - 2007(2) (T = 242, N = 1)
      The dependent variable Y is: LGDP
      The model is:  Y = Trend + Irregular + Cycle 1
      Steady state. found

Log-Likelihood is 1119.67 (-2 LogL = -2239.33).
Prediction error variance is 8.71598e-005

Summary statistics
                      LGDP
T                   242.00
p                   5.0000
std.error           0.0093359
Normality           28.928
H(80)               0.20970
DW                  1.9811
r(1)                0.0090644
q                   19.000
r(q)                -0.031852
Q(q,q-p)            17.710
Rd^2                0.096722

Variances of disturbances:
                    Value           (q-ratio)
Level               0.000000    (   0.0000)
Slope               9.27962e-007 (   0.05981)
Cycle               1.40807e-005 (   0.9075)
Irregular           1.55154e-005 (   1.000)

Cycle other parameters:
```

7.1 Univariate trend-cycle decompositions: GDP

```
Variance               0.00002
Period                20.00000
Period in years        5.00000
Frequency              0.31416
Damping factor         0.60098
Order                  3.00000

State vector analysis at period 2007(2)
                        Value      Prob
Level                 9.35317    [0.00000]
Slope                 0.00670    [0.01262]
Cycle 1 amplitude     0.00326    [      -]
```

The logged GDP series is considered by all three decompositions. A graphical illustration of a decomposition is presented in Figure 7.1.

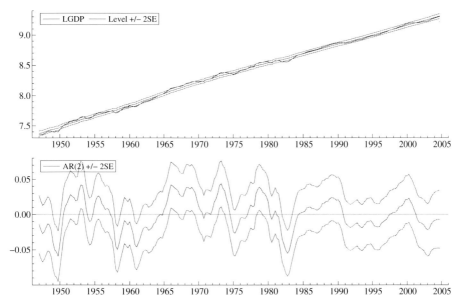

Figure 7.1 US GDP trend-cycle decompositions with a random walk plus fixed drift trend and an AR(2) cycle.

It is argued by Harvey and Koopman (2000) that it is of interest to study the observation weights of the estimated components. Since the models are linear and we assume Gaussian disturbances, the estimated components are linear functions of the observed time series. For example, consider the estimated trend component μ_t at a particular time point $t = \tau$. Observations in the neighboorhood of τ will get more weight than observations more distanced from τ for the estimate of μ_τ. In fact, it can be shown that for so-called time-invariant models, the weights have an exponentially decay as the distance of the observations from τ increases. Also the weight patterns are symmetric in the middle of the time series. However, when the value of τ approaches 1 or T, the weight pattern becomes more and more asymmetric since no observations are available either before time point 1 or after T. Finally it is noted that the weights for μ_τ sum up to one while the weights for the slope, seasonal and irregular components sum up to zero.

The frequency domain equivalence of the weight function is the frequency gain function. It indicates the relative importance of frequencies for the extracted component. In the context of business-cycle tracking, some emphasis to the frequency gain function is given in articles such as Baxter and King (1999) and Christiano and Fitzgerald (2003). The gain function of the estimated cycle component should be band-passed such that only "weight" is given to the business-cycle frequencies (typically the frequencies associated with the range of 2 to 10 years). Harvey and Trimbur (2003) show that in a model-based framework, the estimated cycle component can also possess band-pass filter properties when the cycle specification (4.11) is taken with higher values for k. An illustration of this is given below.

We return to the illustration of the US GDP trend-cycle decomposition based on a smooth trend plus generalised cycle plus irregular model. Go to the **Test/Weight functions** dialog and select the options as:

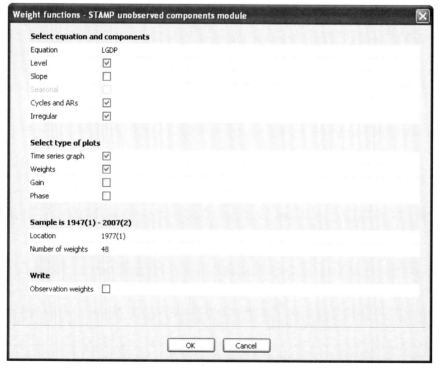

By pressing OK, Figure 7.2 appears. The weights are for the estimated components at time point τ associated with 1977 quarter 1. This is also indicated by the vertical line

7.1 Univariate trend-cycle decompositions: GDP

in the middle of the estimated component plots in first column of graphs. The dialog allows you to choose this position for τ. The smoothed estimate of the trend component is implicitly computed by

$$\widehat{\mu}_\tau = \sum_{j=1-\tau}^{T-\tau} w_j y_{\tau+j},$$

and the weights w_j are presented for index $j = 1-\tau, 2-\tau, \ldots, T-1-\tau, T-\tau$. The trend estimate is rather a smooth function and this is reflected by the distibution of the weights w_j around $j = 0$. The decay to zero is slow (it takes around 6 years !). The decay in the weighting pattern for the cycle component is also slow but the relative importance of the more distanced observations is less compared to those close to τ. Approximately 50% of the weights is determined by the weights of $\tau - 1$, τ and $\tau + 1$. This is even more so for the estimated irregular component. The frequency gain functions are close to a typical decomposition in macroeconomics: the trend is based on the low frequencies, the cycle captures the typical business-cycle frequencies and the high frequencies are for the irregular component.

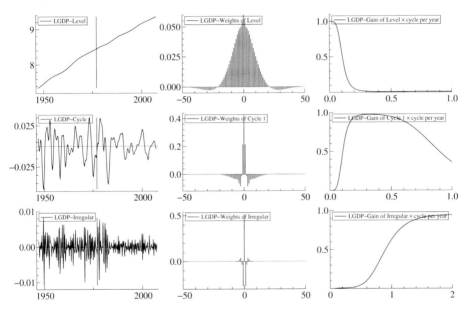

Figure 7.2 Weight and gain functions for the estimated US GDP trend-cycle decomposition based on a smooth trend plus generalised cycle plus irregular model.

Real-time business-cycle tracking focuses on the construction of the filtered estimate of the cycle component. Given that a new observation has just arrived, how do we weight the new observation and the past ones to produce the real-time indicator of the business-cycle ? To obtain some insights, the weight and gain functions can be presented for $\tau = T$. These are shown in Figure 7.3. Further, the different estimates of the cycle component (prediction, filtering and smoothing) can be obtained from the **Test**/More written output dialog by selecting the option 'Cycles and ARs' in the section 'Print recent state values' (expand this section and set value of 'Number of recent periods' to 8). In the Results window, the output is given by

```
Final state values of Cycle 1
         Coef(|t-1)    Coef(|t)    Coef(|T)   Rmse(|t-1)   Rmse(|t)   Rmse(|T)
2006(1)   0.003146    0.003376    0.006210    0.01458    0.01458    0.01038
2006(2)   0.004205    0.004176    0.005976    0.01458    0.01458    0.01086
2006(3)   0.003766    0.003617    0.003452    0.01458    0.01458    0.01158
2006(4)   0.0009131   0.0009323   0.0001707   0.01458    0.01458    0.01269
2007(1)  -0.0007951  -0.0009163  -0.002540    0.01458    0.01458    0.01396
2007(2)  -0.003155   -0.003023   -0.003023    0.01458    0.01458    0.01458
```

The results show that revision of estimates has taken place in the periods before 2007(2) but these revisions are not significant given the standard errors in the range of 0.01 and 0.015. It should also be noted that the business-cycle fluctuations in the years 2006 and 2007 are not significant, perhaps with the exception of the smoothed estimate for 2006(1).

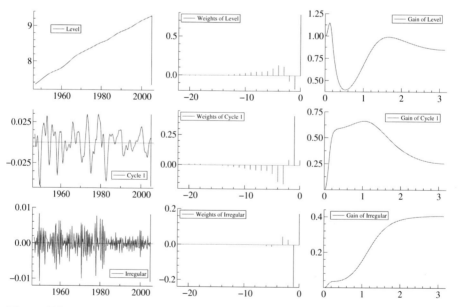

Figure 7.3 Weight and gain functions for the estimated US GDP trend-cycle decomposition based on a smooth trend plus generalised cycle plus irregular model (end-of-sample estimates: filtering).

7.2 Multivariate trends and cycles: GDP and Investment

It is shown in the previous section that US GDP can be decomposed into a trend and a cycle by setting the level variance to be 'Fixed' (though sometimes it is estimated as zero even if it is not constrained at the outset). Using the USmacro07 data over the full period gives (not forgetting to deselect the 'Seasonal' component) a plausible cycle, albeit one that is rather noisy as the irregular variance is estimated to be zero. However, it is not easy to get a good fit with a higher order cycle using this series, see previous section. Another solution is to estimate GDP jointly with Investment as the latter has relatively larger cyclical movements. As shown below, a model with balanced slopes and a fourth-order cycle works well.

To set up the model, proceed to the Formulate a model dialog and select LGDP and LINV as the two dependent variables (label them both as Y). Then proceed to the Select components dialog, deselect 'Seasonal', select 'Cycle short' in 'Cycle(s)' and change 'Order of cycle' from 1 to 4. Also ensure that 'Multivariate settings' is marked in 'Options'. Press OK. In the Select variance matrices and components for each equation dialog, change the entry for the level component of LGDP from 'in' to 'fix'. By this change, the variance associated with the level disturbance of LGDP is fixed at zero. This leads to a smooth trend component for LGDP while the trend for LINV is specified as in a local linear trend model. The slope component is balanced by having its disturbance variance matrix as a scaled variance matrix of ones. This is achieved by changing the 'Var matrix' entry of 'slope' from 'full' to 'ones'. Press OK.

After estimation, the Results window reveals the following output.

```
Level disturbance variance   for LINV: 0.000125089
Level disturbance factor variance for LINV: 0.000125089
Level disturbance factor loading  for LGDP: 0
                LGDP          LINV
Constant        9.357         0.0000

Slope disturbance variance matrix of ones, scaled:
          LGDP          LINV
LGDP   1.226e-006    1.226e-006
LINV   1.226e-006    1.226e-006

Cycle disturbance variance/correlation matrix:
          LGDP          LINV
LGDP   1.733e-005     0.8944
LINV   8.030e-005     0.0004650

Irregular disturbance variance/correlation matrix:
          LGDP          LINV
```

```
LGDP    1.523e-005      0.6962
LINV    6.574e-005      0.0005854

Cycle other parameters:
Period                  31.30762
Period in years         7.82690
Frequency               0.20069
Damping factor          0.47241
Order                   4.00000

State vector analysis at period 2007(2)
Equation LGDP
                        Value       Prob
Level                   9.35689     [0.00000]
Slope                   0.00763     [0.00300]
Cycle 1 amplitude       0.00659     [  .NaN]

Equation LINV
                        Value       Prob
Level                   7.55961     [0.00000]
Slope                   0.00939     [0.00057]
Cycle 1 amplitude       0.04895     [  .NaN]
```

As can be seen the cycles are strongly correlated (0.89). The slopes are balanced and therefore perfectly correlated. However, there is a non-zero deterministic difference between them which is why the actual estimates of the slopes for LGDP and LINV at the end of the sample, 0.0076 and 0.0094 respectively, are not the same. The forecasts will therefore diverge from each other. Note that the model is co-integrated of order (2,1). It is clear from figure 7.4 that the series do not have a single common trend and the role of the stochastic level in the LINV equation is to allow for changes in the long-run proportion of GDP that is taken up by Investment.

7.3 Inflation

7.3.1 Expected inflation

There are many reasons for wanting to estimate the expected rate of inflation. At the most basic level it is important to have a good estimate of the underlying rate of inflation for policy purposes. Governments will typically estimate this as the percentage change in the price level over the past year. However, as shown in Harvey (1989, p. 363), this estimator is inefficient. What is needed is the filtered estimator at the end of the series.

To illustrate, we consider the US inflation quarterly time series from 1947(1)-

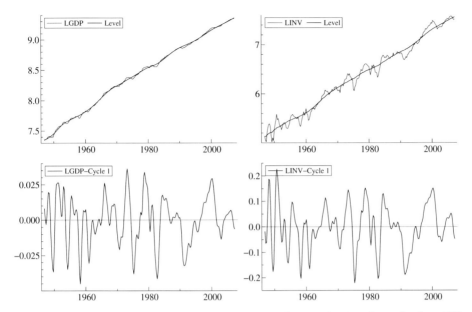

Figure 7.4 Trends and (fourth-order) cycles in LGDP and LINV from database USmacro07.

2007(2) (in logs, ×100) from the database file 'USmacro07.in7'. The inflation series 'INFLcpi' is modelled as a local level model and is estimated as in §4.2.2. Part of the standard output in the Results window is:

```
Variances of disturbances:
                  Value      (q-ratio)
Level           1.85233    (   0.8066)
Irregular       2.29655    (   1.000)

State vector analysis at period 2007(2)
          Value      Prob
Level    5.29714   [0.00001]
```

STAMP can plot and save estimates of the current level of inflation and the predicted level at all points in the series. These can be obtained from the dialog **Test/Component graphics** by first selecting 'Level' in the section 'Select components'. De-select the other components. Next, expand section 'Prediction, filtering and smoothing', select 'Predictive filtering' and 'Filtering' but de-activate 'Estimates in different plots'. The last action ensures that the estimated components are presented in one graph and not (as is the default option) in different graphs. Press OK and Figure 7.5 should appear. It is interesting to compare the estimates from (contemporaneous) filtering and predictive filtering with the smoothed estimates.

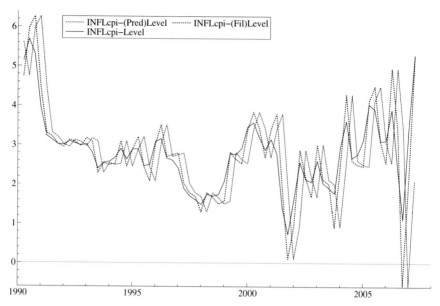

Figure 7.5 Predicted, filtered and smoothed estimates of quarterly inflation (from 1990 onwards).

7.3.2 Inflation and the output gap

Univariate time series models consisting of a random walk plus a stationary component as in the previous section are often used to model inflation. Indeed, Cogley and Sargent (2007, section 2) argue that 'A consensus has emerged that trend inflation is well approximated by a driftless random walk'. The role of the random walk component is to capture the underlying level of inflation. Since its current expected value is the long-run forecast it satisfies the usual definition of core inflation; see Bryan and Cecchetti (1994). The difference between inflation and core inflation is sometimes called the *inflation gap*. Figure 7.6 shows the smoothed components from a UC model for the annualized rate of inflation (INFLcpi) provided in the USMACRO07 data set. A stochastic cycle and a seasonal have been added to a local level model estimated in the previous section. If the inflation gap is estimated by the cycle it is somewhat smoother than the detrended series because the irregular has been filtered out.

The inflation gap can be related to the output gap obtained for US GDP in section 7.1. Figure 7.7 shows them plotted together (with inflation divided by 100). There is clearly some co-movement, but the graph suggests that insisting on a model with time invariant dynamics may be unwise. Output leads inflation in the 1970s, primarily at the time of the two oil crises, but this is not the case later on.

The STAMP program can be used to model the relationship between inflation and output in two different ways. The first is to treat the output gap as an exogenous variable

7.3 Inflation

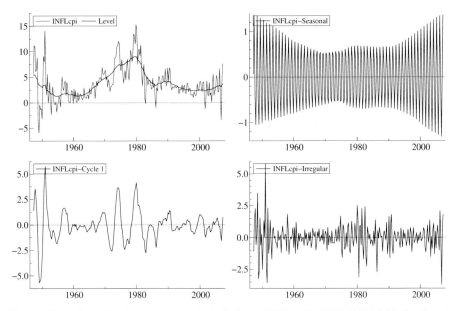

Figure 7.6 Smoothed components in inflation (INFLcpi in USMACRO07 database).

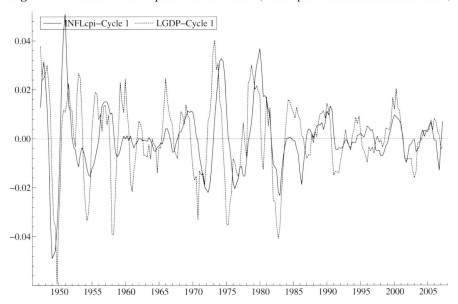

Figure 7.7 Smoothed estimates of the cycles obtained from univariate models for LGDP and INFLcpi in USMACRO07 database.

in a single model. A simple version of the Phillips curve relationship is

$$\pi_t = \mu_t + \gamma_t + \beta x_t + \varepsilon_t, \quad \varepsilon_t \sim NID\left(0, \sigma_\varepsilon^2\right), \quad t = 1, ..., T \tag{7.1}$$

where π_t is inflation, μ_t is the random walk, γ_t is the seasonal component, x_t is a measure of the output gap such as the displayed LGDP cycle in Figure 7.7. The output gap x_t can be lagged. In fact a better fit is obtained over the full sample with x_{t-1}. It is also the case that x_{t-4} is significant but this turns out to be due to the 1970s and is not reproduced later.

The diagnostics are not satisfactory, but given the erratic movements in the 1970s and the subsequent sharp fall in the early 1980s this is not surprising. However, if we start in 1986(1), the model still passes the tests with flying colours. There is no evidence for lags beyond one. The output gap of lag 1 seems to give the best fit, the estimate of β being 44.5 with a t-statistic of 1.96. Figure 7.8, produced from the dialog **Test/Components** graphics is informative in that it shows the underlying level and the effect of the output gap on the level.

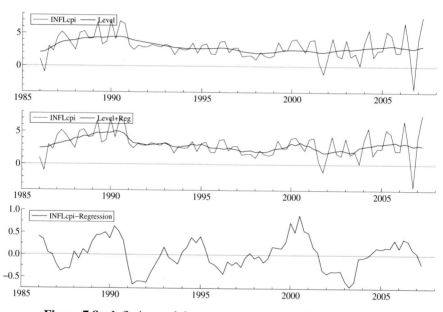

Figure 7.8 Inflation and the output gap as an explanatory variable.

```
Variances of disturbances:
                Value     (q-ratio)
Level         0.122330   (  0.08105)
Seasonal      0.0183967  (  0.01219)
Irregular     1.50934    (  1.000)

State vector analysis at period 2007(2)
                          Value      Prob
Level                   3.10086   [0.00000]
```

```
Seasonal chi2 test          32.22345 [0.00000]

Equation INFLcpi: regression effect in final state at time 2007(2)
           Coefficient          RMSE      t-value       Prob
LGDP-Cycle 1_1   44.50634    22.70598    1.96012  [0.05342]
```

Multi-step forecasts from the end of 1997 are shown in Figure 7.9. They were obtained by going to 'Prediction graphics', moving the radio button to 'Multi-step ahead' and setting 'Post-sample size' to 38. The movements, which are conditional on the output gap, are not big but the higher inflation around 2000 is picked up. The volatility of the series in recent years has made accurate forecasting of any one quarter difficult.

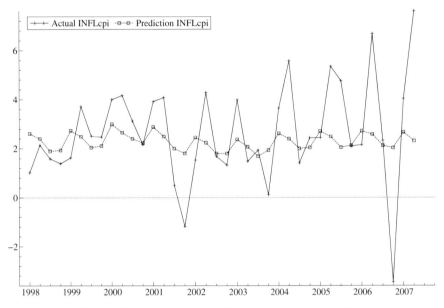

Figure 7.9 Multi-step predictions from end of 1997.

7.3.3 Bivariate modeling

Rather than first estimating the output gap from a univariate model for GDP, inflation and GDP may be modeled jointly as

$$\begin{bmatrix} \pi_t \\ y_t \end{bmatrix} = \begin{bmatrix} \mu_t^\pi \\ \mu_t^y \end{bmatrix} + \begin{bmatrix} \psi_t^\pi \\ \psi_t^y \end{bmatrix} + \begin{bmatrix} \varepsilon_t^\pi \\ \varepsilon_t^y \end{bmatrix} \qquad (7.2)$$

where μ_t^π is a random walk and μ_t^y is an integrated random walk. These two stochastic trends are independent of each other. A seasonal component can be added to the model and in the estimates reported seasonal effects were included in the equation for inflation.

The stochastic cycles are modeled as similar cycles. The paper by Harvey, Trimbur and van Dijk (2007) reports estimates of a model that has the same form as (7.2) except that both trends are integrated random walks. STAMP 8 allows the random walk and integrated random walk restrictions to be imposed on the model.

A simple transformation of the similar cycle model allows the cycle in inflation to be broken down into two independent parts, one of which depends on the GDP cycle, that is $\psi_t^\pi = \beta \psi_t^y + \psi_t^{\pi\dagger}$, where $\beta = Cov(\psi_t^\pi, \psi_t^y) = Cov(\kappa_t^\pi, \kappa_t^y)$. Substituting in the inflation equation in (7.2) gives

$$\pi_t = \mu_t^\pi + \beta \psi_t^y + \psi_t^{\pi\dagger} + \varepsilon_t^\pi \qquad (7.3)$$

If the cycle disturbances κ_t^π and κ_t^y are perfectly correlated, this corresponds to (7.1) if ψ_t^y is set to x_t. However, model (7.1) could itself be extended to include a stochastic cycle.

The model estimated over the full period works quite well - but the case for a contemporaneous relationship is undermined by the lag structure identified from Figure 7.7. With data from 1986(1), the diagnostics are much better. But more to the point, a contemporaneous relationship is reasonable.

To set up the model as outlined above, proceed to the Formulate a model dialog and select INFLcpi first and LGDP second as the two dependent variables (label them both as Y). Then proceed to the Select components dialog, accept the default setting but also select 'Cycle short' in 'Cycle(s)' and keep 'Order of cycle' to 1. Also ensure that 'Multivariate settings' is marked in 'Options'. Press OK. In the Select variance matrices and components for each equation dialog, change the entries as follows.

7.3 Inflation

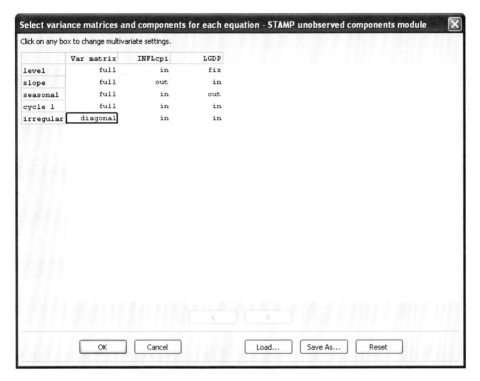

The estimation results are given by

```
Estimation done by Maximum Likelihood (exact score)
The databased used is USmacro07.in7
The selection sample is: 1986(1) - 2007(2) (T = 86, N = 2)
The dependent vector Y contains variables:
        INFLcpi        LGDP
The model is:  Y = Trend + Seasonal + Irregular + Cycle 1
Component selection: 0=out, 1=in, 2=dependent, 3=fix
            INFLcpi        LGDP
Level         1              3
Slope         0              1
Seasonal      1              0
Cycle         1              1
Irregular     1              1
Steady state. found
```

```
Log-Likelihood is 409.231 (-2 LogL = -818.461).
Prediction error variance/correlation matrix is
          INFLcpi        LGDP
```

```
INFLcpi      2.36221     0.06736
LGDP         0.00048     0.00002
```

Summary statistics

	INFLcpi	LGDP
T	86.000	86.000
p	6.0000	6.0000
std.error	1.5369	0.0046125
Normality	10.436	0.015023
H(28)	1.2445	0.80649
DW	1.6742	2.0429
r(1)	0.12071	-0.030650
q	14.000	14.000
r(q)	0.086068	-0.017939
Q(q,q-p)	13.882	15.587
Rs^2	0.36832	0.11933

Variances of disturbances in Eq INFLcpi:

	Value	(q-ratio)
Level	0.0279077	(0.01768)
Seasonal	0.0171006	(0.01083)
Cycle	0.0369929	(0.02344)
Irregular	1.57852	(1.000)

Variances of disturbances in Eq LGDP:

	Value	(q-ratio)
Slope	3.09023e-007	(0.2295)
Cycle	0.000000	(0.0000)
Irregular	1.34651e-006	(1.000)

Level disturbance variance for INFLcpi: 0.0279077
Level disturbance factor variance for INFLcpi: 0.0279077
Level disturbance factor loading for LGDP: 0

	INFLcpi	LGDP
Constant	0.0000	9.346

Slope disturbance variance for LGDP: 3.09023e-007

Seasonal disturbance variance for INFLcpi: 0.0171006

Cycle disturbance variance/correlation matrix:

7.3 Inflation

```
             INFLcpi         LGDP
INFLcpi      0.03699         1.000
LGDP         0.0006950       1.306e-005
```

Irregular disturbance diagonal variance matrix:
```
             INFLcpi         LGDP
INFLcpi      1.579           0.0000
LGDP         0.0000          1.347e-006
```

Cycle other parameters:
```
Period              29.09415
Period in years      7.27354
Frequency            0.21596
Damping factor       0.95921
Order                1.00000
```

Cycle variance/correlation matrix:
```
             INFLcpi         LGDP
INFLcpi      0.4629          1.000
LGDP         0.008697        0.0001634
```

State vector analysis at period 2007(2)
Equation INFLcpi
```
                         Value        Prob
Level                    2.79756   [0.00000]
Seasonal chi2 test      30.83048   [0.00000]
Cycle 1 amplitude        0.34816   [   .NaN]
```

Equation LGDP
```
                         Value        Prob
Level                    9.34643   [0.00000]
Slope                    0.00610   [0.00011]
Cycle 1 amplitude        0.00654   [   .NaN]
```

If there are no restrictions on the irregular, the correlation is minus one. A diagonal variance matrix for the irregular is imposed but it makes virtually no difference to the goodness of fit.

The correlation matrix of the cycle gives an estimate of β equal to $8.697/0.1634 = 53.23$. The perfect correlation means that the implied equation for π_t is effectively as in (7.1). The effect of the output gap on core inflation is shown in Figure 7.10.

If a common factor is imposed on the cycle, by setting inflation to be 'dependent' in Multivariate settings, the results is very similar. The cycle covariance matrix is now

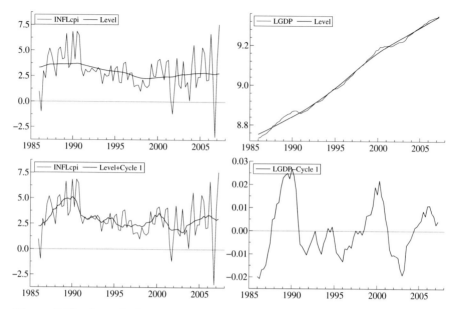

Figure 7.10 Smoothed components from a bivariate model for Inflation and GDP.

as shown below. The factor loading gives the estimate of β.

```
Cycle disturbance variance/correlation matrix:
            INFLcpi         LGDP
INFLcpi     0.03699         1.000
LGDP        0.0006950       1.306e-005
Cycle disturbance factor variance for LGDP: 1.30572e-005
Cycle disturbance factor loading  for INFLcpi: 53.2273
            INFLcpi         LGDP
Constant  -1.665e-016       0.0000
```

The changes in the behaviour of GDP have been less dramatic than for inflation, which suggests using all the GDP data from 1947 and combining it with inflation data from 1986 onwards. The missing observations in inflation can be handled within a state space framework.

Finally we follow Harvey, Trimbur and van Dijk (2007) in fitting second-order cycles. Do this by changing 'Order of cycles' on the Components menu to 2. It will be seen that the second-order cycles are smoother and so may give better measures of the underlying output and inflation gaps. The correlation between them is 0.98 but if the common cycle constraint is imposed it makes very little difference to the fit. (The LR statistic is only 0.08; as shown in Carvalho, Harvey and Trimbur (2007), the 5% critical value is 2.71). The implied value of β is 57, whereas without the common cycle constraint it is 53.

7.4 Stochastic volatility

Let y_t be a stock series returns or the difference of logged exchange rates. Such a series will normally be approximately white noise. However it may not be independent because of serial dependence in the variance. This can be modelled by

$$y_t = \sigma_t \epsilon_t = \sigma \epsilon_t \exp(h_t/2), \quad \epsilon_t \sim \mathsf{IID}(0,1), \quad t = 1,\ldots,\mathsf{T} \tag{7.4}$$

where

$$h_{t+1} = \phi h_t + \eta_t, \quad \eta_t \sim \mathsf{NID}(0, \sigma_\eta^2), \quad |\phi| \leq 1. \tag{7.5}$$

The term σ^2 is a scale factor, ϕ is a parameter, and η_t is a disturbance term which in the simplest model is uncorrelated with ϵ_t; literature reviews are given by Shephard (1996, 2005) and Ghysels, Harvey and Renault (1996). This *stochastic volatility*, (SV) model has two main attractions. The first is that it is the natural (Euler) discrete time analogue of the continuous time model used in papers on option pricing, such as Hull and White (1987). The second is that its statistical properties are easy to determine. The disadvantage with respect to the conditional variance models of the GARCH class is that likelihood based estimation can only be carried out by a computer intensive technique such as that described in Kim, Shephard and Chib (1998) and Sandmann and Koopman (1998). However, a quasi-maximum likelihood (QML) method is relatively easy to apply and is often reasonably efficient. This method is based on transforming the observations to give:

$$\log y_t^2 = \kappa + h_t + \xi_t, \quad t = 1,\ldots,T \tag{7.6}$$

where

$$\xi_t = \log \epsilon_t^2 - E(\log \epsilon_t^2)$$

and

$$\kappa = \log \sigma^2 + E(\log \epsilon_t^2). \tag{7.7}$$

As shown in Harvey, Ruiz and Shephard (1994), the state space form given by equations (7.5) and (7.6) provides the basis for QML estimation via the Kalman filter and also enables smoothed estimates of the variance component, h_t, to be constructed and predictions made. One of the attractions of the QML approach is that it can be applied without the assumption of a particular distribution for ϵ_t.

In Harvey, Ruiz and Shephard (1994), the volatility in the daily exchange rate of the US dollar against four currencies is examined; see also Mahieu and Schotman (1998). The data are in the file EXCH.IN7. After loading EXCH, carry out the transformations of the data for QML estimation. This can be done using the Algebra facility in OxMetrics:

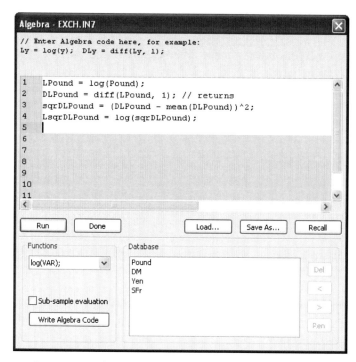

Subsequently you may fit an 'AR(1)' plus 'Irregular' plus 'Level, Fixed' model. Some of the STAMP output is given below for the British pound against the dollar:

```
Summary statistics
  std.error          2.1865
  Normality        181.73
  H(314)             1.0790
  r(1)              -0.033869
  r(29)              0.038984
  DW                 2.0646
  Q(29,27)          13.431
  R^2                0.039738

Variances of disturbances.

Component            Value    (q-ratio)
Level              0.00000  ( 0.0000)
AR(1)              0.73785  ( 0.1592)
Irregular          4.6351   ( 1.0000)

Parameters in AR(1)

Variance           0.73785
AR1 coefficient    0.99598
```

There are a number of things to notice. First, the normality statistic is high. This is inevitable because the transformed model is not Gaussian. It should not worry us.

Second the estimate of ϕ is around 0.996.

The smoothed estimate of the volatility process, h_t, may be extracted in the usual way inside the **Test**/Component graphics dialog. Marking the 'Anti-log analysis' box in the 'Further options' section gives the exponent of the smoothed volatility. This may be interpreted as the ratio of the volatility to the underlying level. It may be preferable to consider the variations in the standard deviation, $\exp(\frac{1}{2}h_{t|T})$, in which case the Calculator in OxMetrics must be used to take the square root; see Figure 7.11.

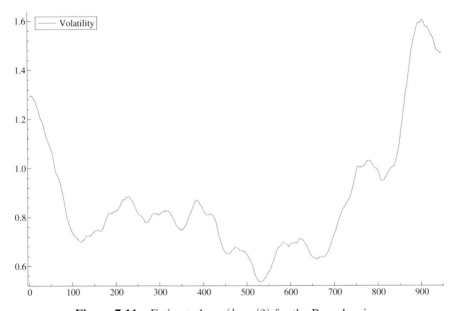

Figure 7.11 Estimated $\exp(h_{t|T}/2)$ for the Pound series.

To estimate σ^2, compute the heteroskedasticity corrected observations:

$$\widetilde{y}_t = y_t \exp(-h_{t|T}/2),$$

and compute the variance $\widetilde{\sigma}^2$. A plot of $\widetilde{\sigma} \exp(\frac{1}{2}h_{t|T})$ against y_t shows how the standard deviation changes with the observations. It may be worth focusing on a shorter period.

7.5 Seasonal adjustment and detrending

One of the attractions of structural time series models is that the trend and seasonal is estimated as part of the overall model. Nevertheless there may be occasions on which the complexity of the full model is such that it cannot be estimated within STAMP. In such circumstances it may be helpful to work with detrended and/or seasonally adjusted data.

7.5.1 Seasonal adjustment

STAMP offers the option of constructing a seasonally adjusted series. This may be saved within the **Test**/Components graphics dialog using the option 'Store selected components in database' or within the **Test**/Store in database dialog. The adjusted series is obtained by extracting the seasonal component in the optimal way from the fitted model. *Thus, given the model specification, it is the best estimator of the non-seasonal part of the series at all time periods, including the beginning and the end.* Of course as more observations become available, the estimates of the seasonal component will change, particularly near the end of the series. This is a natural consequence of model-based seasonal adjustment.

For most purposes, seasonal adjustment based on the basic structural model is recommended. However, it is worth noting that the estimates of the seasonal effects seem, in practice, to be relatively insensitive to the specification of the trend and the inclusion of cycles.

Finally, when working with monthly data, it may sometimes be desirable to allow for calendar effects; see Harvey (1989, pp. 333–7). This may be done by including appropriately formulated explanatory variables in the model. An example of the inclusion of a trading day effect in a structural time series model can be found in Kitagawa and Gersch (1984). The effect of moving festivals — primarily Easter — can be modelled by including a variable which gives the number of days in each month affected by the festival. Other references on the use of structural models in seasonal adjustment include den Butter and Mourik (1990), Maravall (1985) and Harvey (1989, Ch. 6).

7.5.2 Detrending

To illustrate the detrending options in STAMP, consider logged US GDP quarterly time series from 1947-2004 (in logs) from the database file 'USmacro07.in7'. Simple detrending can be carried out within Algebra by fitting a smooth spline using the function smooth_hp():

```
smooth_hp(LGDP, 1600, hp);
DetrendedLGDP = LGDP - hp;
```

The value of $\lambda = 1600$ of the spline function is recommended by Hodrick and Prescott (1980) for quarterly data. It is interesting to note that the so-called Hodrick and Prescott filter can be replicated using the smooth local linear trend model with the signal-to-noise ratio of the slope variance fixed at λ^{-1}; see Harvey and Jaeger (1993). To illustrate this, specify in STAMP a smooth local linear trend model (only select 'Level, Fixed', 'Slope, Stochastic' and 'Irregular') and fix parameter values in the Edit and fix parameter values dialog. Then fix $\sigma_\eta^2 = 0$ and $\sigma_\zeta^2 = 0.000625 = 1/1600$:

7.5 Seasonal adjustment and detrending

The smooth estimates of the level and irregular components are numerically equivalent to the Hodrick and Prescott trend and detrended series, respectively, see Figure 7.12.

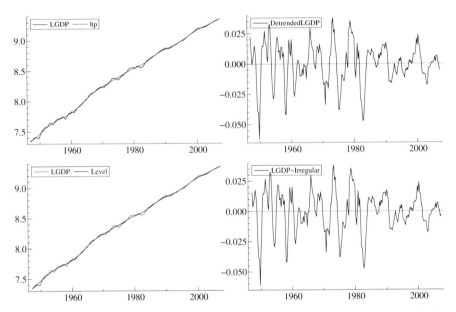

Figure 7.12 Hodrick and Prescott detrending using Algebra in OxMetrics and using local linear trend model in STAMP.

However, detrending with any *ad hoc* procedure should always be used with care because of the danger of creating spurious cycles; see Harvey and Jaeger (1993). A better way to proceed is to fit a model with a trend plus cycle(s) plus irregular, see earlier section. This kind of model-based detrending is quite appealing, though it is important to realize that the properties of the detrended series will not be the same as the properties of the unobserved components in the model which are not part of the trend.

If the series is seasonal, the user may wish to detrend and seasonally adjust. This may be done within STAMP by substrating simultaneously the trend and seasonal unobserved components.

7.6 Missing values

In empirical research, missing observations are often encountered and need to be dealt as part of an econometric time series analysis. The most obvious examples are situations where parts of the time series are not available to the researcher. In other cases, missing observations can occur when, for example, quarterly observations are available for a flow variable until 1990 and monthly observations after 1990. A practical solution is to treat all obervations as monthly and take the first two consecutive months within each three months period as missing until 1990. Further, missing observations can also be systematic when calendar time is the natural time index for a variable that is observed at irregular time intervals. For example, trades in a financial market occur irregular over time but the time-of-day is still relevant for intra-daily seasonality. In this situation, it is preferred to work with a fixed time index where observations are missing at times when no trades occur.

7.6.1 Some missing observations in the time series

We consider the logged electricity consumption (other final users) time series in the database 'ENERGY.IN7' but have treated some observations as missing. The resulting time series with missing values is presented in Figure 7.13. The basic structural time series model is considered for modelling and the estimation output of STAMP is given below.

```
UC( 1) Modelling ofuEL1 by Maximum Likelihood (using ENERGYmiss.in7)
     The selection sample is: 1960(1) - 1986(4)
     with 31 missing observation(s)
     The model is:   Y = Trend + Seasonal + Irregular

Log-Likelihood is 189.831 (-2 LogL = -379.661).
Prediction error variance is 0.00368006

Summary statistics
   std.error          0.060663
   Normality         46.791
   H(24)              0.70541
   r(1)               0.091327
   r(7)               0.082782
   DW                 1.7591
   Q(7,4)             5.6323
   Rs^2               0.94665

Variances of disturbances.

Component                Value     (q-ratio)
Level                 0.00038704  ( 0.7381)
Slope                 1.8795e-006 ( 0.0036)
Seasonal              0.00017306  ( 0.3300)
Irregular             0.00052441  ( 1.0000)
```

7.6 Missing values

```
State vector analysis at period 1986(4)
 - level is 6.34676 with stand.err 0.0469008.
 - slope is 0.0110814 with stand.err 0.00599756.
 - joint seasonal chi2 test is 29.0685 with 3 df.
 - seasonal effects are
period         value       stand.err
     1       0.164786      0.044871
     2      -0.074543      0.051597
     3      -0.152833      0.049351
     4       0.062590      0.045979
```

The output is as usual with the only exception that it is explicitly reported at the beginning that 31 observations were found missing in this analysis. The fact that the output is as usual does support the fact that STAMP can deal with missing observations. To obtain estimates for the missing observations based on this model and on all available data, go to the dialog **Test**/More written output and activate the option 'Missing observation estimates':

```
Missing observation estimates (using full sample)

                  Value        stand.err
1960(1)          5.20814       0.0599084
1960(2)          4.97227       0.0602317
1960(4)          4.98654       0.0673372
1961(4)          5.11206       0.02854
1962(1)          5.46235       0.0400834
1963(2)          5.32565       0.0399539
...
1984(1)          6.3593        0.0393742
1986(2)          6.25005       0.0577701
1986(3)          6.18285       0.0592114
1986(4)          6.40935       0.0622762
```

More evidence of the ability of STAMP to handle missing observations is given by the graphical output. The default graphical output from the dialog Component graphics is presented in Figure 7.14. However, the treatment of missing observations in STAMP becomes most clear when the graphical output of the Weight functions is investigated. The graphs are presented in Figure 7.15 and we learn that weight functions explicitly take account of the missing observations by not giving any weight to missing observations. Time points with missing observations are indicated by black dots in the plots of estimated components. This is done to make it clear at what time points observations are missing. Since the smoothed components are estimated for all time points (interpolation), it needs to be made explicit which observations are missing. The exception is the irregular component from which missing observations can be detected directly since their corresponding estimated irregular values are zero. Other output options in the **Test** menu can be used without problems. The effect that missing observations can have on the statistical output in STAMP can sometimes be quite revealing and interesting.

132   Chapter 7 Applications in Macroeconomics and Finance

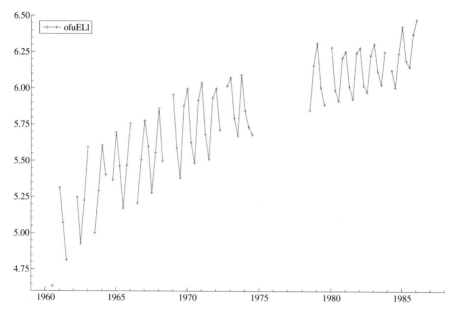

Figure 7.13 Quarterly electricity consumption (other final users) with missing observations.

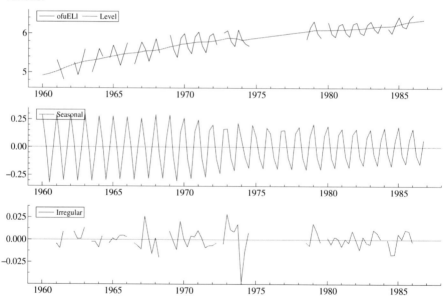

Figure 7.14 Estimates of components when observations are missing.

7.6.2 High-frequency trade prices and many missing observations

High-frequency time series of traded stock prices are usually irregularly spaced. When it is preferred to keep the calendar time and the time stamp of trades is recorded in

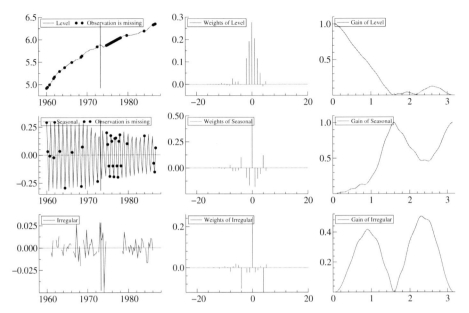

Figure 7.15 Weight and gain functions for components when observations are missing.

seconds, we can have the time index in seconds and treat seconds for which no trades take place as missing. Such a time series (prices in logs) is presented in Figure 7.16 and can be found in the OxMetrics database 'TRADEPRICE.IN7'.

Financial prices are usually modelled as a random walk process. In efficient markets, all relevant information is accounted for in the latest observed price. Since the traded prices in Figure 7.16 are observed at a high frequency, they can be subject to so-called micro-structure noise, see Hansen and Lunde (2006). A basic model to capture the random walk price and the micro-structure noise is the random walk plus noise model (the local level model). Given the irregular spacing of observed prices and the possible wide gaps (in seconds) between observations, we introduce more smoothness to the price process by considering a smooth trend model ('Level, Fixed', 'Slope, Stochastic' and 'Irregular') and estimate this model in STAMP. The deafult output is

```
UC( 1) Modelling LPrice by Maximum Likelihood
    The selection sample is: 1 - 23400
    with 20012 missing observation(s)
    The model is:  Y = Trend + Irregular

Log-Likelihood is 26716.7 (-2 LogL = -53433.4).
Prediction error variance is 5.40634e-008

Summary statistics
  std.error      0.00023252
```

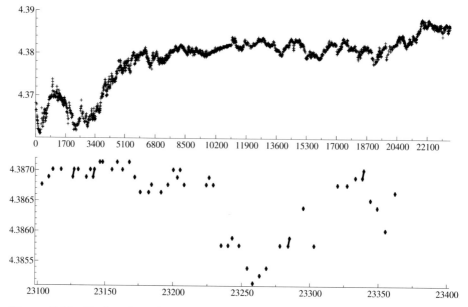

Figure 7.16 Trade prices (logs) for one day and for the last five minutes (second by second).

```
Normality              769.26
H(1128)                0.48463
r(1)                   0.013652
r(57)                 -0.0029337
DW                     1.9718
Q(57,55)             465.03
Rd^2                   0.57049

Variances of disturbances.

Component              Value       (q-ratio)
Level                0.00000      ( 0.0000)
Slope                8.9082e-011  ( 0.0022)
Irregular            3.9721e-008  ( 1.0000)

State vector analysis at period 23400
 - level is 4.38665 with stand.err 0.0016858.
 - slope is 6.77663e-006 with stand.err 6.41375e-005.
```

We note the high number of missing observations in this long time series and the bad diagnostic statistics although the fit is reasonable. These results show that there is much more to the modelling of such time series than this simple model and hence the recent interest in the analysing and modelling of financial high-frequency time series. Nevertheless, it is interesting to show that high-frequency time series with many missing observations can be handled by STAMP. The results of signal extraction is presented in Figure 7.17.

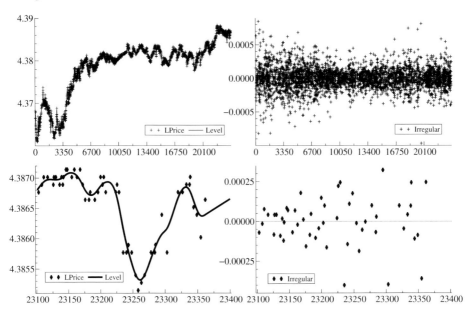

Figure 7.17 Trade price decomposition for one day and for the last five minutes (second by second).

7.7 Exercises

(1) Fit a univariate stochastic volatility model to the DM series in the EXCH.IN7 database. Compare the fit to that recorded above for the pound. Fit a bivariate model for the DM and the pound. What is the quasi-likelihood improvement for the bivariate model over the two univariate models? Is it significant? How might you extract a common cycle from the multivariate model?

(2) Suppose we wish to estimate the underlying rate of inflation with seasonal data. Take first differences of the price level, p, in UKCYP and fit a local level plus seasonal plus irregular. Seasonally adjust the series, store it and make a graphical comparison with the seasonal differences of p divided by 4, that is $\Delta_4 p_t/4$. Does one series lag behind the other? Which method of estimating the deseasonalised rate of inflation do you prefer? See exercise 6.3 of Harvey (1989, p. 363).

(3) Fit a trivariate model to LGDP, LINV and LCONS in 'USmacro07.in7'. You may want to use the data from 1960 onwards and focus on a trend-cycle decomposition.

Chapter 8

Tutorial on Model Building and Testing

This chapter explains how to set up models in STAMP and evaluate the results. First you need to load the data SEATBQ into OxMetrics. This tutorial will be centred around the **Model** window of OxMetrics that is typically started by pressing Alt+y:

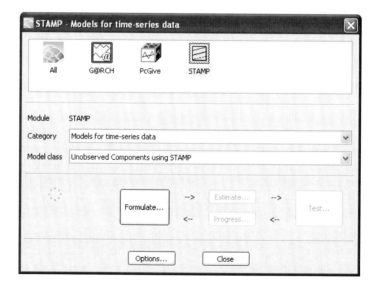

Familiarity with the ideas underlying structural time series modelling is assumed. The aim is to systematically describe the various dialogs rather than to explain the underlying statistical methodology. Those interested in the substance of this application, which concerns the assessment of the effect of the seat belt law in Great Britain, should read Harvey and Durbin (1986).

8.1 Specification of univariate models

To start the process of specifying a model, press `Alt+y` to start the **Model** window. When the STAMP pictogram is not highlighted, click on the 'STAMP' pictogram, select 'Models for time-series data' as 'Category' and 'Unobserved Components using STAMP' as 'Model class'. To start formulate a model, press the 'Formulate' button:

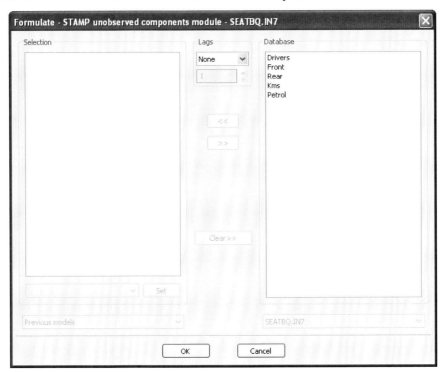

8.1.1 Formulate a model

The Formulate dialog specifies the variables to be included in a model, assigns their status as dependent (Y) or explanatory, and creates lagged values. You can move around the options in the usual way, with the mouse or with `Tab` and `Arrow` keys.

In this seat belt example, the 'Drivers' series will be used as the dependent variable, with Kms and Petrol being explanatory variables.

Select the series from the 'Database' listbox which need to be transferred to the 'Selection' listbox. One way is to click on one database variable in the 'Database' listbox and press the << button. You may also use the multiple selection facility by clicking on database variables and holding the `Ctrl` key simultaneously. Then, press the << button. The first series from the multiple selection will be labelled as dependent (Y). The other variables are not labelled which imply that they are explanatory variables (X). To change the status of a variable in the model, select it and use the droplist below the 'Selection' listbox where you can choose from 'Y variable', 'X variable' and 'Clear status'.

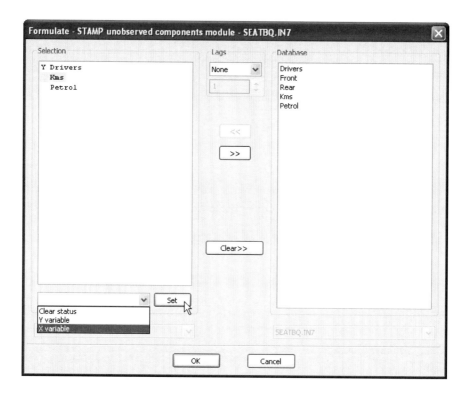

To remove variables from the 'Selection' listbox, select the appropriate variables in this listbox and press the >> button. To create lagged variables, mark the relevant variables in the 'Database' listbox, change the option from 'None' to 'Lag' in the droplist of the 'Lags' section (between the 'Selection' and 'Database' listboxes. Select the appropriate lag number in the editbox below the droplist. Then press the << button. Alternatively, you may also want to select 'Lag 0 to' in the droplist of the 'Lags' section. In this case all lags from 0 to the specified lag will be moved to the 'Selection' listbox.

8.1 Specification of univariate models

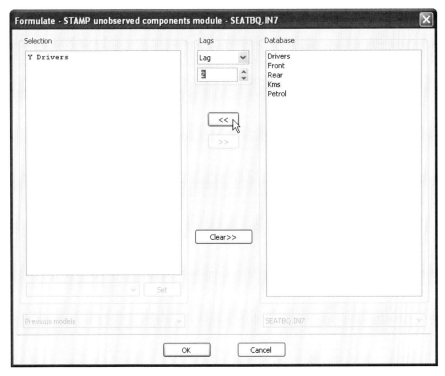

The lagged variables which are not required can be deleted by selecting the variables in the 'Selection' listbox and pressing the >> button.

The model for Drivers does not require lags so if you created them, delete them now.

Two further options in the Formulate dialog are offered by the two droplists at the bottom of the dialog:

- In each STAMP session, previous models are stored in memory and its settings can be reloaded using the 'Previous models' droplist. Various details of the model are reloaded including component specifications, interventions, parameter estimates and restrictions.
- When multiple databases are active in OxMetrics, one can switch between databases by changing the database using the droplist below the 'Database' listbox.

The button Clear>> removes the current model or the model that is formulated. Once this button is pressed, all settings are lost and model formulation can be restarted in this dialog.

The OK button leads you to the Select components dialog.

8.1.2 Selection of components

The Select components dialog:

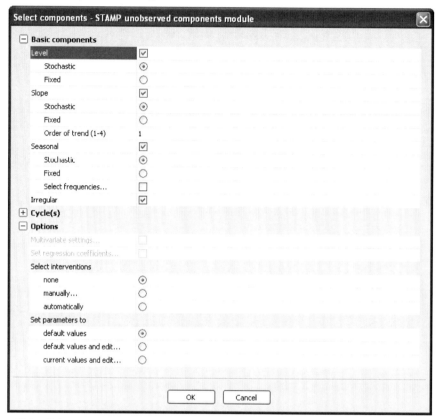

This dialog allows you to specify the unobserved components to be included in the model. The meaning of the different choices should be clear to anyone with a knowledge of structural time series models. If they are not, the tutorial on components, Chapter 4, should be consulted. The combinations needed to specify some of the more common time series models of trend are shown below:

(1) Local level or random walk plus noise – 'Level, Stochastic ' selected and 'Slope' not selected;
(2) Local level with drift – 'Level, Stochastic' and 'Slope, Fixed' selected;
(3) Smooth trend – 'Level, Fixed' and 'Slope, Stochastic' selected;
(4) Generalised trend – 'Level, Fixed' and 'Slope, Stochastic' selected and value 'Order of trend' is 2, 3, or 4;

Note that certain combinations are inadmissible. For example, you cannot have a slope with no level. When 'Level' is not selected but 'Slope' is selected, the program automatically activates the level component as 'Level, Fixed'.

8.1 Specification of univariate models

The default model is the *basic structural model*. To change the setting, move the cursor to the appropriate box in the usual way; that is, use the mouse. Alternatively, you can use the Arrow keys to change options within a group. The Tab key can be used to change from one group of options to another. The first three groups of components are selected by check boxes. The options 'Stochastic' and 'Fixed' can be specified using radio buttons. The slope component is further determined by the 'Order of trend': this value must be an integer and can only take the value 1 (default), 2, 3 and 4. The groups associated with the level and slope and seasonal components belong to the section 'Basic components'.

To select other (stationary) components, the section 'Cycles' need to be expanded by clicking on 'Cycles'. Here, for example, two out of three of the 'Cycle' boxes may be marked together with 'AR(1)' or 'AR(2)' (or even both). The three cycles are distinguished from each other by having different starting values for the period (or frequency) of the cycle. These are starting values for the numerical optimisation procedure. They can be changed in another dialog later.

Having specified the components to enter into the model, there are two ways of proceeding for univariate models. The first possibility is to move directly to the Estimate dialog. This is appropriate if the model is fully specified and can be achieved by keeping the default settings in the 'Options' section ('Select interventions none' and 'Set parameters to default values') and by pressing OK. The Estimate dialog is discussed in §8.2. The second possibility is to choose one or more options in the 'Options' section. When explanatory variables are selected, the option 'Set regression coefficients' is activated by default. This option opens the 'Regression coefficients' dialog and allows the user to select time-varying regression parameters with a choice of different specifications.

The option 'Select interventions manually' opens the Select interventions dialog while the option 'Select interventions automatically' will activate the automatic outlier and break detection procedure during the estimation process. The options 'Set parameter values to default and edit' and 'Set parameter values to current values and edit' open the Edit and fix parameters dialog. In the former case, the default values are displayed while in the latter case, the current parameter values are displayed (when available). You may also want to return to the Formulate dialog which requires pressing Cancel button and type Alt+y (within the OxMetrics program).

After leaving the Select components dialog, it can be re-entered at any time from the Model window (press Alt+y within OxMetrics) and press the Estimate button.

The newly re-entered Select components dialog will always display the current specified model which can be amended at any time.

The set of unobserved components in the model for Drivers should only include the stochastic level, the trigonometric (stochastic) seasonal and the irregular. The slope term should not be selected.

8.1.3 Specify regression coefficients

The Regression coefficients dialog consists of a series of droplist boxes:

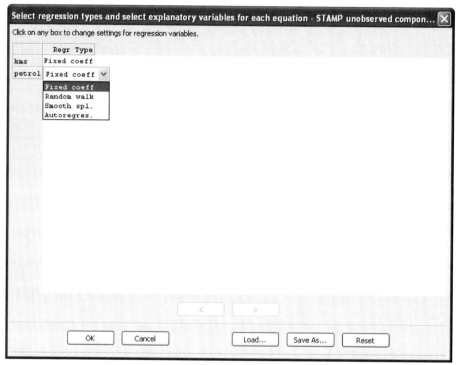

For each explanatory variable, one can select the nature of the regression coefficient. The default option is a fixed parameter for each regression coefficient. However, time-varying regression parameters can also be considered and you can choose three different specifications for a specific time-varying regression coeffient δ_t:

- Random walk: $\delta_t = \delta_{t-1} + u_t$, where $u_t \sim \text{NID}(0, \sigma_u^2)$;
- Smoothing spline : $\Delta \delta_t = \Delta \delta_{t-1} + u_t$, where $u_t \sim \text{NID}(0, \sigma_u^2)$;
- Return to normality : $\delta_t - \delta = \rho_\delta (\delta_{t-1} - \delta) + u_t$, where $u_t \sim \text{NID}(0, \sigma_u^2)$ and δ is the "long-term" fixed coefficient.

The additional parameter σ_u^2 and, possibly, ρ_δ will be simultaneously estimated with the other parameters in STAMP. For the return to normality specification, the additional coefficient δ is part of the fixed regression coefficients and is placed in the state vector.

Once the different specifications are selected for each regression parameter, press OK.

For the Drivers model, the regression coefficients are taken as fixed parameters and the default settings can be accepted.

8.1.4 Selection of interventions

The Select interventions dialog

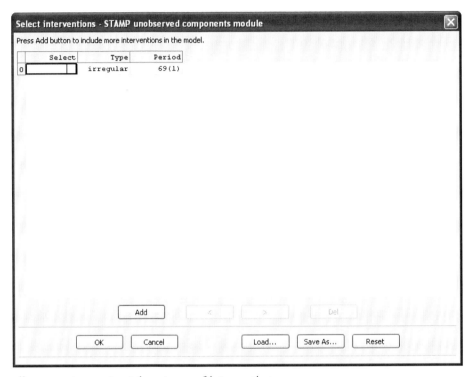

allows you to construct three types of interventions:

- *Impulse/Irregular* - an impulse intervention only affects the observation in question. It can be regarded as picking up an abnormal value of the irregular component, perhaps caused by a measurement error or a strike. In other words the observation in question is regarded as an *outlier*.
- *Level/Step* - a step intervention is a permanent increase in the level of the trend. It represents a structural break, perhaps caused by a change in policy.
- *Slope/Staircase* - a slope intervention leads to a permanent change in the direction of the trend.

All interventions are incorporated as explanatory variables in the model directly. To construct an intervention variable, enter a date in the column 'Period' either by typing the date or by double-clicking on the date and using the mouse and/or arrow keys. The so-called spin keys can be used to increase or decrease the year number or period number. The default type of an intervention is Irregular. By double-clicking on this entry in the 'Type' column, a droplist appears from which you can choose the appropriate intervention: irregular, level and slope. Once the appropriate intervention type and date is entered, the intervention becomes part of the model when the checkbox in the first column 'Select' is activated. A new entry for an intervention appears when the button Add, at the bottom of the dialog, is pressed. In this way a long list of interventions can be created from which interventions can be included or excluded in the model (using the 'Select' column).

The options 'Load' and 'Save As' allow the inclusion of a list of interventions that is created before and saved by the 'Save As' option. The 'Save As' dialog

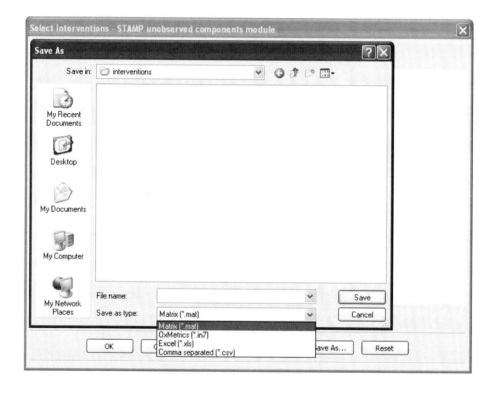

can save the list of interventions in different formats and the 'Load' dialog

8.1 Specification of univariate models

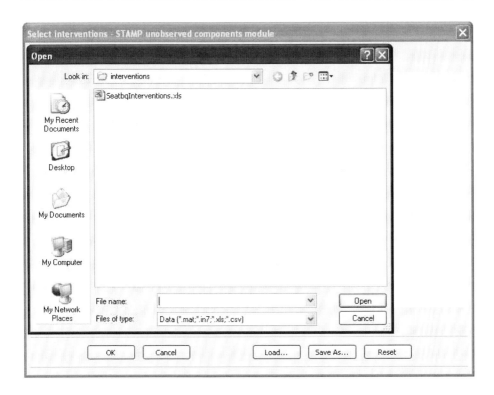

can load a list of interventions in different formats.

The British Government brought in a law at the end of January 1983 which made the wearing of seatbelts compulsory. In our seat belt example we might represent the effect of this law by a level intervention in 83Q1.

8.1.5 Edit and fix parameter values

The dialog Edit and fix parameter values can be accessed when the option 'Edit and fix parameter values' is activated in the Select components dialog:

146 *Chapter 8 Tutorial on Model Building and Testing*

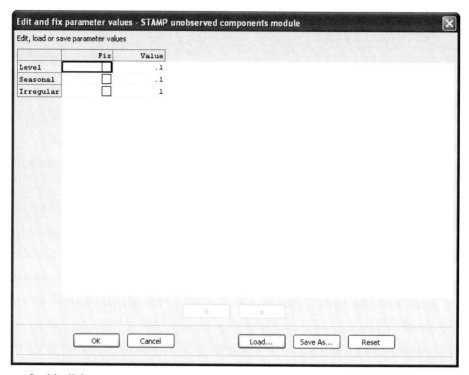

In this dialog you can

- set initial values of parameters that are used for starting the numerical maximisation of the likelihood function in the Estimate dialog, see below.
- set parameter values and fix them at these values so that these will **not** be estimated.

In the entries of the last column 'Value', parameter values can be edited in the usual way. In the first column 'Fix' it is indicated whether parameters need to be fixed at their current values (checkbox is signed) or whether these values are used as starting values for estimation (checkbox is empty). In this dialog, the options 'Load' and 'Save As' are also available to load and save, respectively, parameter value settings.

For the Seatbelt case, just keep the default of estimating all parameter values and press OK.

8.2 Estimate a model

The Estimate Model dialog gives the start sign to the estimation procedure of STAMP. When this has finished, some basic estimation results appear in the Results window of OxMetrics.

8.2.1 Estimate dialog

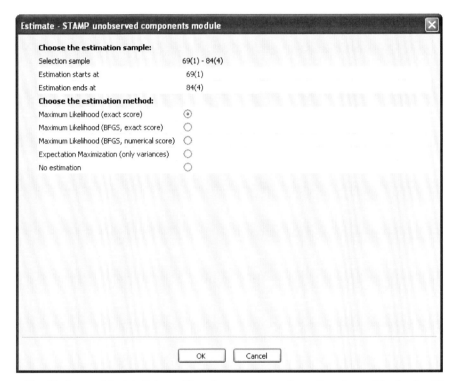

The Estimate Model dialog offers the user a number of options. Firstly the 'Estimation sample' may be set to something other than the 'Selection sample' implied by the database in OxMetrics. Changes are made to the sample period by accessing the appropriate edit boxes using the mouse and inputting the new date. The Estimate Model dialog provides different ways of estimating a structural time series model by the method of maximum likelihood (ML). The different procedures are described in the next section.

8.2.2 Maximum likelihood

The first option 'Maximum Likelihood (exact score)' is the fully automatic procedure which is the one to choose when you are a first user of STAMP.

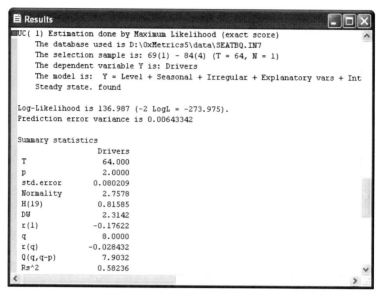

The estimation procedure automatically enters into an initial estimation routines. The details of these routines are given in §9.6. The initial estimation is based primarily on the EM algorithm and a naive nonlinear estimation method. Once the initial estimation routine has obtained a solution close to the ML estimator, the program switches to computing full ML estimates using a numerical optimisation procedure. Again you should refer to §9.6 to understand exactly what is happening here. It should be stressed, however, that STAMP has been designed in such a way that, for most users, the estimation phase can be regarded as a "black box".

Once estimation is in progress, some dots are printed: it confirms that the estimation is in progress. In most situations, the on-line printing of the dots will not be noticed. However, when the time series is very long and/or the model is large, the estimation process may take some time and printing the dots may be useful.

When the optimisation routine has converged or it has reached the maximum number of iterations, it prints some output to the Results window about the model and the estimation procedure. Furthermore, the program produces plots of the estimated components in the Model graphics window. More output and, very importantly, more graphical output can be generated from the dialogs of the **Test** menu; see §8.3.

There are four different estimation methods:

- Maximum likelihood (exact score): the default procedure developed especially for STAMP;
- Maximum likelihood (BFGS, exact score): the standard BFGS method of numerical optimisation where exact scores are computed; it is implemented for the Ox programming language and known as the function MaxBFGS().
- Maximum Likelihood (BFGS, numerical score): as the previous method but

8.2 Estimate a model

based on numerical scores;
- Expectation Maximisation (only variances): the standard EM method using the `Ox/Ssfpack` functions, it is only implemented for the estimation of variances. Other parameters are kept fixed during the estimation method.

Finally, the estimation part can also be circumvented in STAMP. Either by cancelling the dialog (press the red box in the upper-right corner of the dialog) or by selecting 'No estimation' and proceed as normal by pressing OK. In this case, the current (initial) parameter values are taken and output can be generated as detailed below. To emphasize that the model is not estimated, the default output in the Results window is limited to

```
UC( 1) No estimation done
    The databased used is SEATBQ.IN7
    The selection sample is: 69(1) - 84(4) (T = 64, N = 1)
    The dependent variable Y is: Drivers
    The model is:  Y = Level + Seasonal + Irregular + Expl...
    instead of
UC( 1) Estimation done by Maximum Likelihood (exact score)
    The database used is SEATBQ.IN7
    The selection sample is: 69(1) - 84(4) (T = 64, N = 1)
    The dependent variable Y is: Drivers
    The model is:  Y = Level + Seasonal + Irregular + Expl...
    Steady state. found

Log-Likelihood is 136.987 (-2 LogL = -273.975).
Prediction error variance is 0.00643342

Summary statistics
                    Drivers
T                   64.000
p                    2.0000
std.error            0.080209
Normality            2.7578
H(19)                0.81585
DW                   2.3142
r(1)                -0.17622
q                    8.0000
r(q)                -0.028432
Q(q,q-p)             7.9032
Rs^2                 0.58236
```

```
Variances of disturbances:
                    Value        (q-ratio)
Level           0.000350702     (  0.07148)
Seasonal        1.25876e-005    ( 0.002566)
Irregular       0.00490627      (    1.000)
```

8.2.3 Options

In the main Model menu (to be accessed via Alt+y, the button Options gives access to the dialog Options.

Here you can alter the convergence tolerance and change the number of maximum iterations. Reducing the convergence tolerance means that the gradients must be closer to zero for convergence to be deemed to have taken place. A smaller convergence tolerance will usually entail more iterations.

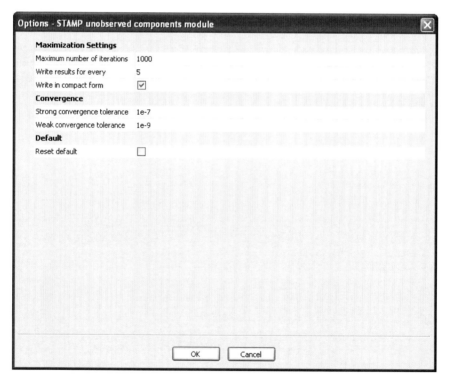

Further you can request more output during estimation by reducing the times that output is written to the Results window and by de-activating the option for 'Write in compact form'.

8.2.4 Progress

In the main Model menu (to be accessed via Alt+y, the button Progress gives access to the the Progress dialog

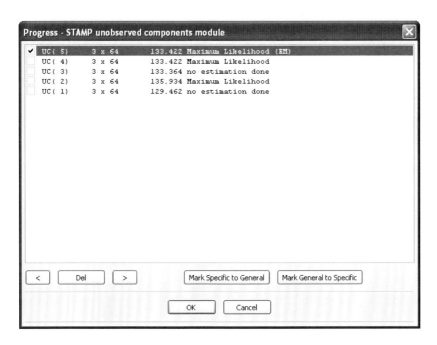

This dialog allows the user to regain the settings of previous models that are considered in STAMP. It also facilitates comparisons between models.

8.3 Model evaluation and testing

Once estimation is complete, STAMP outputs an estimation report, a diagnostic summary report, the estimated variances of the disturbances and a state analysis in the Results window. Further output may be obtained from the Test menu. The Test menu is accessed from the Model menu (press Alt+y) and by pressing the button 'Test'.

152   *Chapter 8 Tutorial on Model Building and Testing*

Each of the eight available options here offers a dialog which enables you to specify exactly what statistics you would like. The first dialog produces written output only, the second dialog is mainly concerned with graphical output of estimated components. The third dialog provides more insights about the signal extraction carried out by STAMP. The fourth to sixth dialogs give graphical diagnostics based on various residual series. They further allow the user to carry out diagnostic checking by written. The seventh dialog offers forecasting facilities. The final eighth dialog gives options to store estimated components in the database of OxMetrics. They are described in the next section.

The output is generated using advanced statistical methods such as the Kalman filter, smoothing algorithms, etc. Most graphics and diagnostic statistics generated from the Test menu are standard and easy to understand. However, if you require some guidance and examples, go to the previous chapters of this manual. The technical details of the algorithms and the formulae of the statistics are given in Part III of this book.

All the written and graphical output is sent to OxMetrics where it may be edited and saved. Thus, for example, you may add comments to the output as you go along or delete any information, perhaps on an unsuccessful run, which you regard as superfluous.

The results shown below apply to the Drivers quarterly seat belt model with a stochastic level, no slope, stochastic seasonal, two explanatory variables (Kms and Petrol), level intervention at 83.1 and irregular. The restriction described for the Parameter vector dialog is not imposed. The model is estimated by Maximum likelihood.

8.3.1 Estimation report

The most important piece of information in the estimation report is the message 'Very strong/Strong/Weak/No convergence in ... iterations' The precise definition of these terms can be found in §9.6, but from a practical point of view, the appearance of the word 'strong' is to be welcomed. If convergence is not strong, you have several possibilities:

- Proceed to examine the model output in order to see how well it fits and where, if anywhere, it appears to deficient.
- Return to the Options dialog, and increase the number of iterations. Re-start the estimation process.
- Formulate a new model.
- Go to the Edit and fix parameter values dialog and start estimation from a new set of initial parameter values. Also some parameter values may be altered and restricted before the estimation process is re-started.

Convergence in the 'Drivers' model is 'Very strong'. The estimation part of the output is given below:

```
Estimating...
Very strong convergence relative to 1e-007
- likelihood cvg 2.07477e-015
- gradient cvg 6.3165e-009
- parameter cvg 2.8362e-008
- number of bad iterations 0
Estimation process completed.

UC( 1) Estimation done by Maximum Likelihood (exact score)
    The database used is SEATBQ.IN7
    The selection sample is: 69(1) - 84(4)  (T = 64, N = 1)
    The dependent variable Y is: Drivers
    The model is:  Y = Level + Seasonal + Irregular + Expl...
    Steady state. found
```

8.3.2 Diagnostic summary report

This output is by default automatically generated after the estimation phase. It presents the basic goodness of fit and diagnostic statistics. Based on this information you can probably decide whether it is worth proceeding to the **Test** menu or whether the model is so poor that you need to reconsider its specification. When convergence is not strong, it may be that iterating further will eventually produce an estimated model which satisfies the diagnostics.

An excessive amount of residual *serial correlation* is a strong indication that the model is not adequately capturing the dynamic structure of the series. The statistics denoted $r(j)$ give the autocorrelation at lag j, while $Q(p,q)$ is the *Box–Ljung* statistic based on the first p autocorrelations; it should be tested against a χ^2 distribution with q degrees of freedom. The classical *Durbin–Watson* test is also given.

The *heteroskedasticity* test statistic, $H(h)$, is the ratio of the squares of the last h residuals to the squares of the first h residuals where h is set to the closest integer of $T/3$. It is centred around unity and should be treated as having an F distribution with (h, h) degrees of freedom. A high (low) value indicates an increase (decrease) in the variance over time.

The *normality* test statistic is the *Bowman–Shenton* statistic based on third and fourth moments of the residuals and having a χ^2 distribution with 2 degrees of freedom when the model is correctly specified. The 5% critical value is thus 5.99. High values are often caused by outliers. The model should not necessarily be rejected, but further investigation, perhaps by looking at the auxiliary residuals, is required.

The basic goodness of fit measure is the *prediction error variance*. Its square root is shown as the equation standard error, 'Std. Error'. The appropriate 'R$\hat{2}$' (R^2) measure is also given; the statistics R_d^2 and R_s^2 are more suitable in time series as they compare the fit with a random walk plus drift and a random walk with fixed seasonal dummies, respectively.

Part of the default output is also concerned with the estimated state vector. This output will be discussed in more detail below.

Again the Drivers model is given below as an illustration.

```
UC( 1) Estimation done by Maximum Likelihood (exact score)
       The database used is D:\OxMetrics6\data\SEATBQ.IN7
       The selection sample is: 69(1) - 84(4) (T = 64, N = 1)
       The dependent variable Y is: Drivers
       The model is:  Y = Level + Seasonal + Irregular + Expl...
       Steady state. found

Log-Likelihood is 136.987 (-2 LogL = -273.975).
Prediction error variance is 0.00643342

Summary statistics
                    Drivers
T                   64.000
p                   2.0000
std.error           0.080209
Normality           2.7578
H(19)               0.81585
DW                  2.3142
r(1)               -0.17622
q                   8.0000
r(q)               -0.028432
Q(q,q-p)            7.9032
Rs^2                0.58236

Variances of disturbances:
                    Value      (q-ratio)
Level           0.000350702  (  0.07148)
```

8.3 Model evaluation and testing

```
Seasonal     1.25876e-005  ( 0.002566)
Irregular    0.00490627    (    1.000)

State vector analysis at period 84(4)
                           Value       Prob
Level                      4.65232  [0.00830]
Seasonal chi2 test        70.32562  [0.00000]
Seasonal effects:
             Period        Value       Prob
                  1      -0.07388  [0.00692]
                  2      -0.14215  [0.00001]
                  3      -0.01376  [0.62260]
                  4       0.22979  [0.00000]

Regression effects in final state at time 84(4)

                  Coefficient      RMSE    t-value       Prob
Level break 83(1)    -0.21887    0.05351   -4.09053  [0.00014]
Kms                   0.22469    0.17688    1.27031  [0.20913]
Petrol               -0.26824    0.12211   -2.19675  [0.03212]
```

8.3.3 More written output

The first dialog in the **Test** menu is the More written output dialog.

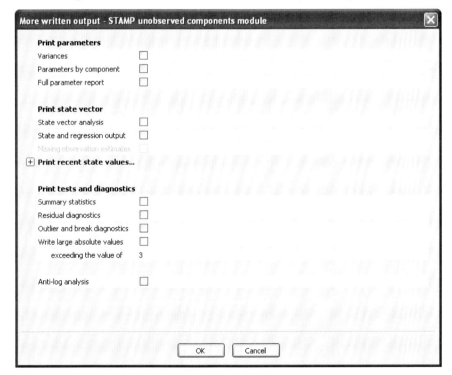

It provides a variety of further information for the estimated model. In the 'Print parameters' section it reports on the parameters (usually estimated by maximum likelihood): variances, all parameters organised by each component and the vector of parameters including the transformations that are used during the estimation process and standard errors. These options print estimated standard deviations and estimated parameters associated with the cycle (damping factors and periods) and the autoregressive components. The estimated standard deviations of the disturbances in the model are printed. The figures in parentheses, the *q-ratios*, are the ratios of each standard deviation to the standard deviation associated with the largest variance.

```
Standard deviations of disturbances:
                 Value       (q-ratio)
Level           0.0187270   (  0.2674)
Seasonal        0.00354790  (  0.05065)
Irregular       0.0700448   (  1.000)

Full parameter report
Actual parameters (all)
                 Value
Var Level       0.00035070
Var Seasonal    1.2588e-005
Var Irregular   0.0049063
Transformed parameters (not fixed)
                 Transform      1stDer        2ndDer      asymp.s.e
Var Level       -3.9778      -2.3137e-008   -0.12331     0.36506
Var Seasonal    -5.6414      -1.3678e-008   -0.019777    0.94585
Var Irregular   -2.6586       2.6068e-008   -1.3692      0.11596
Actual parameters (not fixed) with 68% asymmetric confidence interval
                 Value        leftbound     rightbound
Var Level       0.00035070   0.00016899    0.00072783
Var Seasonal    1.2588e-005  1.8984e-006   8.3464e-005
Var Irregular   0.0049063    0.0038908     0.0061868
```

The section 'Print state vector' in the Written more output dialog contains options that print values taken by the various components at the end of the sample. This information is stored in what we call the *state* vector. The default output from the 'State and regression output' option has the following particular feature. The figure in square brackets after the *t-value* is a two-sided Prob. value which shows the probability of getting an absolute value of a standard normal variable greater than the *t-value* if the true parameter is zero. Additional output is generated for the seasonal and cycle components when it is relevant to the model. The generated output is discussed in Chapter 10.

In many cases the data are in logs and it is useful to activate the check box 'Anti-log analysis'. This provides additional information. For example, the actual value of the level is given and the seasonals can be interpreted as the factors by which to multiply the trend. Finally, the default option to 'Get steady state' is discussed in Chapter 9 and it is advised to keep this option activated.

8.3 Model evaluation and testing

The regression output for the explanatory variables and the interventions are given below.

```
State vector anti-log analysis at period 84(4)
It is assumed that time series is in logs.
                           Value        Prob
Level (anti-log)        104.82788  [0.00830]
Level (bias corrected)  445.49849  [   .NaN]
Seasonal chi2 test       70.32562  [0.00000]
Seasonal effects:
                Period      Value        Prob      %Effect
                  1       0.92878  [0.00692]      -7.12178
                  2       0.86749  [0.00001]     -13.25105
                  3       0.98633  [0.62260]      -1.36666
                  4       1.25834  [0.00000]      25.83404

State vector at period 84(4)
             Coefficient      RMSE       t-value        Prob
Level           4.65232      1.70110      2.73489  [0.00830]
Seasonal        0.18597      0.02605      7.14005  [0.00000]
Seasonal 2     -0.03006      0.02181     -1.37805  [0.17357]
Seasonal 3      0.04382      0.01731      2.53148  [0.01414]

Regression effects in final state at time 84(4)
                    Coefficient      RMSE       t-value        Prob
Level break 83(1)     -0.21887      0.05351     -4.09053  [0.00014]
Kms                    0.22469      0.17688      1.27031  [0.20913]
Petrol                -0.26824      0.12211     -2.19675  [0.03212]
```

Note that in our example 'Kms' is not statistically significant but 'Petrol' is significant.

The intervention estimate is -0.22, implying a fall in the level of front seat passengers killed and seriously injured of $100(1 - \exp(-.22))$ which is around 20%.

The option 'Print recent state values' allows the user to investigate the different estimates (prediction, filtering and smoothing) of the different components in the recent period. This information can be useful to assess the level of revisions in estimates. For the level and seasonal components, the output is given by

	Coef(|t-1)	Coef(|t)	Coef(|T)	Rmse(|t-1)	Rmse(|t)	Rmse(|T)
83(4)	4.857	5.001	4.633	1.750	1.744	1.698
84(1)	5.001	4.888	4.640	1.744	1.736	1.699
84(2)	4.888	4.886	4.645	1.736	1.736	1.699
84(3)	4.886	4.839	4.650	1.736	1.735	1.700
84(4)	4.839	4.652	4.652	1.735	1.701	1.701

Final state values of Seasonal

	Coef(|t-1)	Coef(|t)	Coef(|T)	Rmse(|t-1)	Rmse(|t)	Rmse(|T)
83(4)	0.2370	0.2254	0.2296	0.03474	0.03257	0.03013
84(1)	-0.07647	-0.07053	-0.07368	0.02646	0.02498	0.02444

158 Chapter 8 Tutorial on Model Building and Testing

```
84(2)     -0.1361     -0.1353     -0.1420     0.02908     0.02780     0.02708
84(3)     -0.01802    -0.01005    -0.01368    0.02897     0.02774     0.02692
84(4)      0.2225      0.2298      0.2298     0.03398     0.03122     0.03122
```

The last section 'Print tests and diagnostics' outputs options that can also be outputted from other dialogs, see below.

8.3.4 Components graphics

The information provided here is fundamental to the interpretation of the model. The smoothed estimates of the components are obtained using all the information in the sample; that is, they are constructed using observations which come after as well as those which come before. This is sometimes referred to as *signal extraction*. The Components graphics dialog controls the way this information is presented. In particular it provides options to graph particular components and combinations of components. The use of the 'Zoom sample range' option is particularly useful here as it allows an enormous amount of information to be very quickly assimilated.

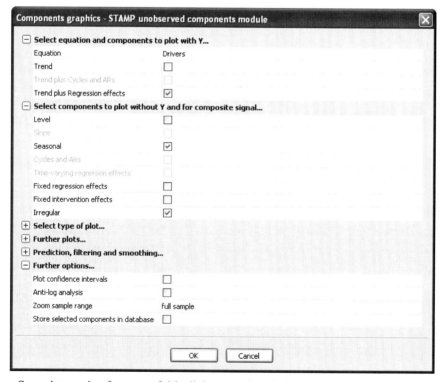

Some interesting features of this dialog are given below.

- All components included in the model can be plotted together with the sum of regression effects ('Fixed regression effects') and the sum of intervention effects

8.3 Model evaluation and testing

('Fixed intervention effects'). Confidence intervals can be graphed when the option 'Plot confidence intervals' is activated.
- The 'Detrended Y' and 'Seasonally adjusted Y' options in the 'Further plots' section allow subtraction of the estimated trend and seasonal component, respectively, from a particular series. These options often show some interesting features of the data which may not be clear at first sight.
- Each seasonal effect j, for $j = 1, \ldots, s$, can be graphed against the years of the data-set. This set of s (the number of seasonals) graphs is obtained by marking the checkbox 'Individual seasonals' in the 'Further plots' section.
- When the data is recorded in logs, the components may be transformed by taking anti-logs (exp) which can be requested in the 'Further options' section using 'Anti-log analysis'.
- Sometimes it is more useful to focus on the predicted and/or filtered estimates; that is, the one-step ahead predictions and/or the concurrent estimates of the components, respectively. These options are available in the section 'Prediction, filtering and smoothing'.
- All series generated through this dialog can be saved in the database of OxMetrics by pressing the button Store select components in database.
- The signal extraction sample can be changed in this dialog using option 'Signal extraction sample'. This option is different than 'Zoom sample range'. The latter option has no effect on the sample selection, it just focuses on a specific sample range in the plots.

8.3.5 Weight functions

To gain insights in how the estimated components are computed, weight functions can be quite useful. Estimation is based on appropriate weighting of observations. The weights are imposed by the estimated model, see Harvey and Koopman (2000). The frequency domain counterparts of the weighting function are the spectral gain function and the phase. All these functions can be requested from the Weight functions dialog for each estimated component and for each time period in the sample.

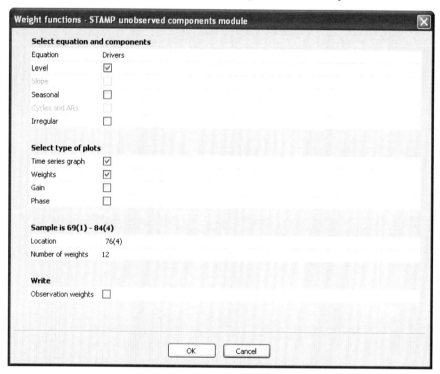

It may be interesting to view how the weight and gain functions change when the 'Location' in the sample approaches the end of the sample.

8.3.6 Residuals graphics

The basic way of assessing the suitability of a time series model is by examining its (standardised) residuals. Based on the true values of the parameters, they are independent and normally distributed with zero mean and unit variance. Diagnostic checking of the model may be carried out by plotting the residuals against time and checking the *Correlogram* to see whether the autocorrelations are small. A graph of the distribution of the residuals can also be constructed. An alternative way of examining the residuals is by looking at their frequency domain properties using the *Spectrum*.

The Residuals graphics dialog, with the default options marked, is

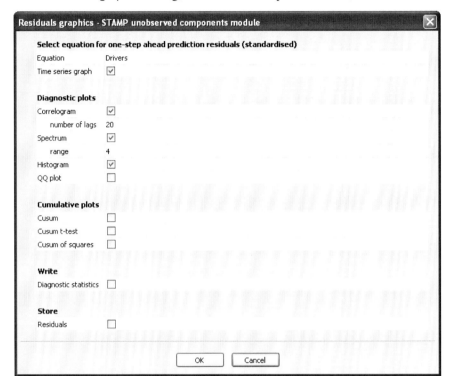

The graph options in the Residuals graphics dialog can be requested in the usual way. The diagnostics checking tools are discussed in §10.6 which also gives some appropriate definitions and formulae. The option 'Write diagnostic tests' sends some serial correlation statistics and normality tests to the OxMetrics Results window. The Store section allows you to save the residuals (which are sometimes called the *innovations*).

8.3.7 Auxiliary residuals graphics

The auxiliary residuals are smoothed estimates of the irregular, level and slope disturbances. Although they are neither serially uncorrelated nor uncorrelated with each other, they play a valuable role in that they go some way towards separating out pieces of information which are mixed up together in the innovation residuals. In particular they are helpful in detecting and distinguishing between outliers and structural change.

In the 'Drivers' case there is some indication of a downward shift in the levels of the series after the oil crisis in the mid-1970s. Thus there may be a role for additional level intervention variables. Can you comment on the relation between the level residuals of

'Drivers'?

![Auxiliary residuals graphics - STAMP unobserved components module dialog box showing: Select equation and auxiliary residuals (t-tests for)... with Equation Drivers; checkboxes for Irregular (outlier intervention) ☑, Level (break in level intervention) ☑, Slope (break in slope intervention) ☐ (greyed out); Select plots section with Index plot ☑, Histogram ☑, QQ plot ☐; Write section with Normality tests ☐, Large absolute values ☐ exceeding the value of 3; Store section with Selected auxiliary residuals ☐; OK and Cancel buttons]

The options of the Auxiliary residuals graphics dialog can be requested in the usual way. The first three check boxes indicate which auxiliary residuals should be subjected to the requested diagnostics. A particularly useful option is 'Large absolute values .. exceeding the value of' in the 'Write' section which prints information of large values in specific auxiliary residuals. This includes the actual value, which must be larger than 3.0 (default) in absolute values, and the corresponding date. The normality tests, which are adjusted for serial correlation, are printed in the Results window when the checkbox 'Normality tests' in the 'Write' section is marked.

8.3.8 Prediction testing

This dialog is for making predictions over periods at the end of the sample for which we have taken out the most recent observations from the estimation sample. Note that the predictions are therefore made within the parameter estimation sample. The formulae of the related statistics are discussed in §10.8.

One-step ahead predictions can be made and their consistency with the predictions in the sample period can be assessed by looking at graphs of their predictive perfor-

8.3 Model evaluation and testing

mance and examining statistics such as the *predictive failure test* which are produced when 'Prediction tests' option is marked in the 'Write' section. The statistics are reported in the Results window. The default settings of the Prediction graphics dialog is given below.

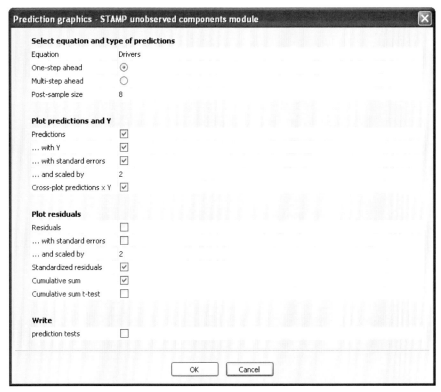

When 'Multi-step ahead' is chosen, the predictions are made using the information at the end of the sample period minus the post-sample size but the predictions are not updated in the post-sample with the arrival of new observations. In other words they are extrapolations or multi-step predictions.

8.4 Forecasting

The Forecasting command in the **Test** menu leads to the Forecasting dialog.

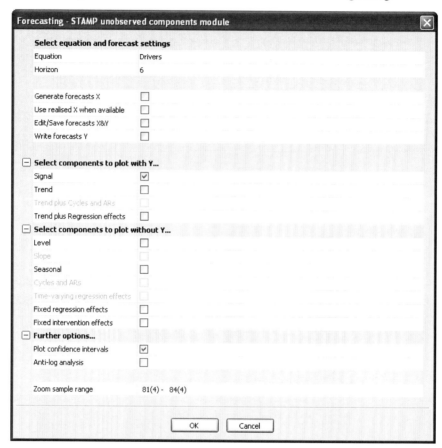

8.4.1 Without explanatory variables

Forecasting without explanatory variables is straightforward. As well as the series itself, the individual components may be forecasted by marking the appropriate boxes. Graphs of forecasts together with a prediction interval of 68% (that is, one RMSE on either side) can be produced. The 'Anti-log analysis' option takes the anti-log (adjusted as decribed in §10.5). The forecasts are printed in the Results window when 'Write forecasts Y' is marked. The length of the forecast horizon can be altered using the first edit field in this dialog.

To use the forecasts later for another model, the forecasts may be edited and saved to a file by marking 'Edit/Save forecasts X'. This gives entrance to a matrix editor in

8.4 Forecasting

which the forecasts (implied by the model) can be edited and can be saved. Note that when the forecasts are changed in the Matrix Editor, STAMP will keep its forecasts unaltered.

8.4.2 Interventions

If the interventions have been created using the Intervention dialog, appropriate future values are created automatically.

8.4.3 Explanatory variables

When forecasts are to be made conditional on future values of an explanatory variable, the default is to set these values equal to the last value in the sample. However, you may want to alter the default values in four different ways using the option 'Generate forecasts X' in the 'Forecast settings' section:

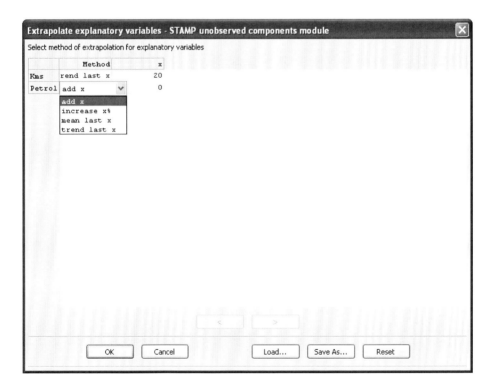

(1) All explanatory variables can be extrapolated by a pre-specified method for each explanatory variable and altering the associated edit field appropriately. If the data are in logarithms the incrementation corresponds to a growth rate. The following methods are available:

- add x: for every post-sample period, value of x is added, consecutively.
- increase $x\%$: for every post-sample period, value is increased by percentage x, consecutively.
- mean last x: values for post-sample period is extrapolated by the mean of the last x in-sample observations.
- trend last x: values for post-sample period is extrapolated by the trend of the last x in-sample observations.

(2) The option 'Edit/Save forecasts X&Y' can be used to edit the extrapolated values for each X variable.

(3) When the end date of the estimation sample is before the end date of the database, the available observations of the explanatory variables can be used by choosing the option 'Use realised X when available'. The graph will also present the realised Y.

(4) Finally, the option 'Edit/Save forecasts X&Y' activates the Matrix Editor and allows editing the forecasts of explanatory variables. Such forecasts can be generated by separate models for the explanatory models. Their forecasts can be written in the Results window by the option 'Write forecasts Y':

Forecasts with 68% confidence interval from period 84(4) forwards:

	Forecast	stand.err	leftbound	rightbound
1	7.14066	0.08518	7.05547	7.22584
2	7.07239	0.08904	6.98335	7.16142
3	7.20078	0.09065	7.11013	7.29142
4	7.44433	0.09052	7.35382	7.53485
5	7.14066	0.09359	7.04706	7.23425
6	7.07239	0.09711	6.97527	7.16950

When the end date of the estimation sample is before the end date of the database, the Results window will also output forecast accuracy measures such as the root mean squared (percentage) error (RMSE) and the mean absolute (percentage) error (MAPE).

Part III

Statistical Treatment

Chapter 9

Statistical Treatment of Models

To understand fully the output of STAMP it is useful to understand the way in which the models are handled statistically. The algorithms used to carry out the computations are based on the state space form. The interested reader may find explanations of the underlying ideas and proofs of the algorithms in Anderson and Moore (1979), Harvey (1989, Chs. 3 and 4), de Jong (1991) and Koopman (1993). This chapter simply sets out the algorithms and gives the exact technical details of implementation. We first formally define the models which were used in Chapters 4–6.

9.1 Model definitions

9.1.1 Univariate time series models

A univariate model may be written as

$$y_t = \mu_t + \gamma_t + \psi_t + \nu_t + \epsilon_t, \quad \epsilon_t \sim \text{NID}(0, \sigma_\epsilon^2), \quad t = 1, \ldots, T, \qquad (9.1)$$

where μ_t is the trend, γ_t is the seasonal, ψ_t is the cycle, ν_t is a first-order autoregressive component and ϵ_t is the irregular. The model can be extended with two similar cycle components; see below.

The stochastic trend component is specified as

$$\begin{aligned} \mu_t &= \mu_{t-1} + \beta_{t-1} + \eta_t, & \eta_t &\sim \text{NID}(0, \sigma_\eta^2), \\ \beta_t &= \beta_{t-1} + \zeta_t, & \zeta_t &\sim \text{NID}(0, \sigma_\zeta^2), \end{aligned}$$

where β_t is the slope or gradient of the trend μ_t. The irregular ϵ_t, the level disturbance η_t and the slope disturbance ζ_t are mutually uncorrelated. The slope component β_t can be excluded from the trend specification when this is appropriate. Some special trend specifications are listed in Table 9.1.

The seasonal component has the trigonometric seasonal form and is given by

$$\gamma_t = \sum_{j=1}^{[s/2]} \gamma_{j,t}$$

169

Table 9.1 Some special level and trend specifications.

Level	σ_ϵ	σ_η	
constant term	*	0	
local level (LL)	*	*	
random walk (RW)	0	*	

Trend	σ_ϵ	σ_η	σ_ζ
deterministic	*	0	0
LL with fixed slope	*	*	0
RW with fixed drift	0	*	0
local linear (LLT)	*	*	*
smooth trend	*	0	*
second differencing	0	0	*
Hodrick-Prescott	*	0	$.025\,\sigma_\epsilon$

Asterix * indicates any positive value.

where each $\gamma_{j,t}$ is generated by

$$\begin{bmatrix} \gamma_{j,t} \\ \gamma_{j,t}^* \end{bmatrix} = \begin{bmatrix} \cos\lambda_j & \sin\lambda_j \\ -\sin\lambda_j & \cos\lambda_j \end{bmatrix} \begin{bmatrix} \gamma_{j,t-1} \\ \gamma_{j,t-1}^* \end{bmatrix} + \begin{bmatrix} \omega_{j,t} \\ \omega_{j,t}^* \end{bmatrix}, \quad \begin{array}{l} j=1,\ldots,[s/2], \\ t=1,\ldots,T, \end{array}$$

where $\lambda_j = 2\pi j/s$ is the frequency, in radians, and the seasonal disturbances ω_t and ω_t^* are two mutually uncorrelated NID disturbances with zero mean and common variance σ_ω^2. For s even, the component at $j = s/2$ collapses to

$$\gamma_{j,t} = \gamma_{j,t-1}\cos\lambda_j + \omega_{j,t}.$$

The statistical specification of a cycle, ψ_t, is given by

$$\begin{bmatrix} \psi_t \\ \psi_t^* \end{bmatrix} = \rho_\psi \begin{bmatrix} \cos\lambda_c & \sin\lambda_c \\ -\sin\lambda_c & \cos\lambda_c \end{bmatrix} \begin{bmatrix} \psi_{t-1} \\ \psi_{t-1}^* \end{bmatrix} + \begin{bmatrix} \kappa_t \\ \kappa_t^* \end{bmatrix}, \quad t=1,\ldots,T,$$

where ρ_ψ, in the range $0 < \rho_\psi \leq 1$, is a *damping factor*; λ_c is the frequency, in radians, in the range $0 \leq \lambda_c \leq \pi$; κ_t and κ_t^* are two mutually uncorrelated NID disturbances with zero mean and common variance σ_κ^2. Note that the *period* of the cycle is equal to $2\pi/\lambda_c$. There may be two additional cycles of this form incorporated in the model.

A first-order autoregressive, AR(1), process is given by

$$\nu_t = \rho_\nu \nu_{t-1} + \xi_t, \quad \xi_t \sim \text{NID}(0,\sigma_\xi^2),$$

with ρ_ν in the range $0 < \rho_\nu < 1$. The AR(1) component is actually a limiting case of the stochastic cycle when λ_c is equal to 0 or π, though it is specified separately in STAMP, partly to avoid confounding and partly because it is not a limiting case in multivariate models.

Finally, the disturbances driving each of the components in the model are mutually uncorrelated.

9.1.2 Explanatory variables and interventions

A single equation model may include exogenous explanatory variables, lagged values of the dependent variable and intervention variables, as well as unobserved components such as trend, seasonal and cycle. Thus (9.1) can be extended as

$$y_t = \mu_t + \gamma_t + \psi_t + r_t + \sum_{\tau=1}^{p} \phi_\tau y_{t-\tau} + \sum_{i=1}^{k}\sum_{\tau=0}^{q} \Delta_{i\tau} x_{i,t-\tau} + \sum_{j=1}^{h} \lambda_j w_{j,t} + \varepsilon_t \quad (9.2)$$

where x_{it} is an exogenous variable, w_{jt} is an intervention (dummy) variable and $\phi_\tau, \Delta_{i\tau}$ and λ_j are unknown parameters.

9.1.3 Multivariate time series models

Multivariate models have a similar form to univariate models, except that $\mathbf{y}_t$ is now an $N \times 1$ vector of observations, which depends on unobserved components which are also vectors. Thus in the special case of a multivariate local level model

$$\begin{aligned} \mathbf{y}_t &= \boldsymbol{\mu}_t + \boldsymbol{\epsilon}_t, & \boldsymbol{\epsilon}_t &\sim \text{NID}(\mathbf{0}, \boldsymbol{\Sigma}_\epsilon), \\ \boldsymbol{\mu}_t &= \boldsymbol{\mu}_{t-1} + \boldsymbol{\eta}_t, & \boldsymbol{\eta}_t &\sim \text{NID}(\mathbf{0}, \boldsymbol{\Sigma}_\eta), \end{aligned} \quad (9.3)$$

where $\boldsymbol{\Sigma}_\epsilon$ and $\boldsymbol{\Sigma}_\eta$ are both $N \times N$ variance matrices, and $\boldsymbol{\eta}_t$ and $\boldsymbol{\epsilon}_t$ are mutually uncorrelated in all time periods. The other disturbances in the general model (9.1) similarly become vectors which have $N \times N$ variance matrices. Models of this kind are called *seemingly unrelated time series equations* (SUTSE).

The other components are incorporated in a multivariate model in a similar manner by letting the components and disturbances become $N \times 1$ vectors and letting the variances become $N \times N$ variance matrices. In the case of the cycle, the parameters ρ and λ are the same for all series. As regards the disturbances,

$$E\left(\boldsymbol{\kappa}_t \boldsymbol{\kappa}_t'\right) = E\left(\boldsymbol{\kappa}_t^* \boldsymbol{\kappa}_t^{*'}\right) = \boldsymbol{\Sigma}_\kappa \text{ and } E\left(\boldsymbol{\kappa}_t \boldsymbol{\kappa}_t^{*'}\right) = \mathbf{0}. \quad (9.4)$$

The specification of the cycle model can also be written as

$$\begin{bmatrix} \boldsymbol{\psi}_t \\ \boldsymbol{\psi}_t^* \end{bmatrix} = \left\{ \rho_\psi \begin{bmatrix} \cos \lambda_c & \sin \lambda_c \\ -\sin \lambda_c & \cos \lambda_c \end{bmatrix} \otimes \mathbf{I_N} \right\} \begin{bmatrix} \boldsymbol{\psi}_{t-1} \\ \boldsymbol{\psi}_{t-1}^* \end{bmatrix} + \begin{bmatrix} \boldsymbol{\kappa}_t \\ \boldsymbol{\kappa}_t^* \end{bmatrix}, \quad (9.5)$$

where

$$Var \begin{bmatrix} \boldsymbol{\kappa}_t \\ \boldsymbol{\kappa}_t^* \end{bmatrix} = \mathbf{I_2} \otimes \boldsymbol{\Sigma}_\kappa, \quad \boldsymbol{\psi}_t \text{ and } \boldsymbol{\psi}_{t-1}^* \text{are } \mathbf{N \times 1} \text{ vectors.}$$

A stationary first-order vector autoregressive may be included in a multivariate model as an alternative to, or even as well as, a cycle. Thus

$$\nu_t = \Phi \nu_{t-1} + \xi_t, \quad \text{and} \quad \text{Var}(\xi_t) = \Sigma_\xi, \tag{9.6}$$

where Φ is a $N \times N$ matrix of coefficients. The condition for ν_t to be stationary is that the roots of the matrix polynomial $\mathbf{I} - \Phi \mathbf{L}$ should all lie outside the unit circle.

9.1.4 Common factors

In a common factor model, some or all of the components are driven by disturbance vectors with less than N elements. In terms of a SUTSE model, the presence of common factors means that the variance matrices of the relevant disturbances are less than full rank.

Common factors may appear in all components, including the irregular. Common trends arise through common levels, common slopes or a combination of the two.

9.1.4.1 Common levels

Consider the local level model, (9.3), but suppose that the rank of Σ_η is $K < N$. The model then contains K common levels or common trends and may be written as

$$\begin{aligned} y_t &= \Theta \mu_t^\dagger + \mu_\theta + \epsilon_t, & \epsilon_t &\sim \text{NID}(0, \Sigma_\epsilon), \\ \mu_t^\dagger &= \mu_{t-1}^\dagger + \eta_t^\dagger, & \eta_t^\dagger &\sim \text{NID}(0, \mathbf{D}_\eta), \end{aligned} \tag{9.7}$$

where $\eta_t^\dagger$ is a $K \times 1$ vector, Θ is a $(N \times K)$ matrix of *standardised factor loadings*, $\mathbf{D}_\eta$ is a diagonal matrix and μ_θ is a $(N \times 1)$ vector in which the first $N - K$ elements are zeros and the last K elements are contained in a vector $\overline{\mu}$. The standardised factor loading matrix contains ones in the 'diagonal' positions, i.e. $\theta_{ii} = 1, i = 1, \ldots, K$, while $\theta_{ij} = 0$ for $j > i$.

The model may be recast in the original SUTSE form (9.3) by writing $\mu_t = \Theta \mu_t^\dagger + \mu_\theta$ and noting that $\Sigma_\eta = \Theta \mathbf{D}_\eta \Theta'$ is a singular matrix of rank K. When there are no common trends so $N = K$, the factor loading matrix is the Cholesky decomposition of Σ_η. As regards $\overline{\mu}$, partition $\Theta = (\Theta_1', \Theta_2')'$ where Θ_1 consists of the first K rows, and partition μ_t similarly. Then

$$\overline{\mu} = -\Theta_2 \Theta_1^{-1} \mu_{1t} + \mu_{2t}, \quad t = 1, \ldots, T. \tag{9.8}$$

The vector $\overline{\mu}$ is estimated from the above set of equations using the estimated states at $t = T$. The computations are carried out by back-substitution since the load matrix is lower triangular.

9.1.4.2 Smooth trends with common slopes

Common slopes may be formulated along similar lines. We first consider the case where the variance matrix of the slope disturbances is of rank K_β but the variance matrix of levels is null so that the estimated trends are relatively smooth. The model is therefore

$$\begin{aligned} y_t &= \mu_t + \epsilon_t, & \epsilon_t &\sim \text{NID}(0, \Sigma_\epsilon), \\ \mu_t &= \mu_{t-1} + \beta_{t-1}, \\ \beta_t &= \beta_{t-1} + \zeta_t, & \zeta_t &\sim \text{NID}(0, \Sigma_\zeta), \end{aligned}$$

but, following (9.7) above, it may be re-formulated as

$$\begin{aligned} y_t &= \Theta\mu_t^\dagger + \mu_{\theta t} + \epsilon_t, & \epsilon_t &\sim \text{NID}(0, \Sigma_\epsilon), \\ \mu_t^\dagger &= \mu_{t-1}^\dagger + \beta_{t-1}^\dagger, \\ \beta_t^\dagger &= \beta_{t-1}^\dagger + +\zeta_t^\dagger, & \zeta_t^\dagger &\sim \text{NID}(0, \mathbf{D}_\zeta), \end{aligned}$$

where $\Sigma_\zeta = \Theta \mathbf{D} \Theta'$ and the first K_β elements in $\mu_{\theta t}$ are zeros and the remainder are contained in a vector $\overline{\mu} + \overline{\beta}t$. The $(N-K) \times 1$ vectors, $\overline{\mu}$ and $\overline{\beta}$, may both be calculated by expressions like (9.8).

9.1.4.3 Common trends: level and slopes

A general multivariate local linear trend model in which the level variance matrix is of rank K_η while the slope variance matrix is of rank K_β may be written

$$\begin{aligned} y_t &= \mu_t + \epsilon_t, & \epsilon_t &\sim \text{NID}(0, \Sigma_\epsilon), \\ \mu_t &= \mu_{t-1} + \Theta_\beta \beta_{t-1}^\dagger + \beta_\theta + \eta_t, & \eta_t &\sim \text{NID}(0, \Sigma_\eta), \\ \beta_t^\dagger &= \beta_{t-1}^\dagger + \zeta_t^\dagger, & \zeta_t^\dagger &\sim \text{NID}(0, \mathbf{D}_\zeta), \end{aligned}$$

where the $N \times K_\beta$ matrix Θ_β is such that $\Sigma_\zeta = \Theta_\beta \mathbf{D}_\zeta \Theta_\beta'$, and $\beta_\theta = (0', \overline{\beta}')'$ with $\overline{\beta}$ a vector of length $N - K_\beta$.

If $K_\beta = 1$, setting Θ_β equal to a vector of ones and letting $\overline{\beta} = 0$ would imply that all series had the same underlying growth rate (when modelling in logs). This might be plausible even if there are no common levels. The implication is that the trends in the forecast function remain parallel, in other words the long-run growth paths are the same. However, unless there are similar restrictions on the levels, the growth paths within the sample will not necessarily have kept together.

The following interpretation can be made if $K_\beta \leq K_\mu$ and the common stochastic slopes only affect the observations *via* the common stochastic trends:

$$\begin{aligned} y_t &= \Theta_\mu \mu_t^\dagger + \mu_{\theta t} + \epsilon_t, & \epsilon_t &\sim \text{NID}(0, \Sigma_\epsilon), \\ \mu_t^\dagger &= \mu_{t-1}^\dagger + \Theta_{1\mu}^{-1} \Theta_{\mu\beta} \beta_{t-1}^\dagger + \eta_t^\dagger, & \eta_t^\dagger &\sim \text{NID}(0, \mathbf{D}_\eta), \\ \beta_t^\dagger &= \beta_{t-1}^\dagger + \zeta_t^\dagger, & \zeta_t^\dagger &\sim \text{NID}(0, \mathbf{D}_\zeta), \end{aligned}$$

where $\Theta_{1\mu}$ is the first K_μ rows of Θ_μ, $\Theta_{\mu\beta}$ is the first K_μ rows of Θ_β, $\mu_{\theta t} = (0', \beta_1' t, \overline{\mu}' + \beta_2' t)'$ with $\overline{\beta}_1$ is a vector of length $K_\mu - K_\beta$, $\overline{\beta}_2$ is a vector of length $N - K_\mu$ and $\overline{\beta} = (\overline{\beta}_1', \overline{\beta}_2')'$. The fixed components satisfy, for any $1 \leq t \leq T$,

$$\overline{\mu} = -\Theta_{2\mu}\Theta_{1\mu}^{-1}\mu_{1t} + \mu_{2t} - \overline{\beta}_2 t \quad \text{and} \quad \overline{\beta} = -\Theta_{2\beta}\Theta_{1\beta}^{-1}\beta_{1t} + \beta_{2t}$$

where $\Theta_{1\beta}$ is the first K_β rows of Θ_β and $\Theta_{2\beta}$ is the last $N - K_\mu$ rows.

At present STAMP does not offer the option of making the restrictions implied by the above formulation. However, if $K_\mu = K_\beta = 1$, it must be the case that the above restriction implies $\Theta_{2\beta} = \Theta_{2\mu}$. More generally for $K_\mu = K_\beta$ we have

$$\mu_t^\dagger = \mu_{t-1}^\dagger + \Theta_{1\mu}^{-1}\Theta_{1\beta}\beta_{t-1}^\dagger + \eta_t^\dagger, \quad \eta_t^\dagger \sim \text{NID}(0, \mathbf{D}_\eta)$$

and setting $\Theta_{1\beta} = \Theta_{1\mu}$ implies the further restriction that the slopes in the common trends are mutually independent. Thus the common trends are fully independent in both level and slope disturbances.

9.1.4.4 Common seasonals

Common factors in seasonality implies a reduction in the number of disturbances driving changes in the seasonal patterns. It does not imply any similarity in seasonal patterns unless the deterministic seasonal components outside the common seasonals are set to zero. Thus suppose, for simplicity, that the series contain only seasonals and irregular. Then

$$y_t = \Theta\gamma_t^\dagger + \gamma_{\theta t} + \epsilon_t,$$

where the last $N - K$ elements of $\gamma_{\theta t}$ contain fixed seasonal effects.

9.1.4.5 Common cycles

A model with common trends and common cycles could be written

$$y_t = \Theta_\mu \mu_t + \mu_\theta + \Theta_\psi \psi_t + \epsilon_t,$$

where Θ_ψ is $N \times K_\psi$. Since the expectation of a cycle is zero, it is not necessary to include a vector of constant terms corresponding to the vector μ_θ which is needed for common levels.

As with other components, the common cycle can be re-formulated as a SUTSE model in which the disturbance variance matrices are $\Sigma_\kappa = \Theta_\psi \mathbf{D}_\kappa \Theta_\psi'$. This is only possible because ρ_ψ and λ are the same in all the common cycles.

9.1.5 Explanatory variables

Explanatory variables and interventions may be included in multivariate models. Thus (9.2) generalises to

$$y_t = \mu_t + \gamma_t + \psi_t + r_t + \sum_{\tau=1}^{r} \Phi_\tau y_{t-\tau} + \sum_{\tau=0}^{s} \delta_\tau x_{t-\tau} + \Lambda w_t + \epsilon_t, \quad t = 1, \ldots, T,$$

where x_t is a $K \times 1$ vector of explanatory variables and w_t is a $K^* \times 1$ vector of interventions. Elements in the parameters matrices, Φ, δ_τ and Λ may be specified to be zero, thereby excluding certain variables from particular equations. In addition, the unobserved components may be subject to common factor restrictions as described in the previous subsection.

9.2 State space form

The statistical treatment of structural time series models is based on the state space form (SSF). We use a particular SSF which is close to the 'state space model' of de Jong (1991). It joins a *measurement* equation

$$y_t = Z_t \alpha_t + X_t b + G_t u_t, \quad t = 1, \ldots, T, \tag{9.9}$$

with a *transition equation*, which allows the *states* α_t to evolve according to a first-order vector autoregressive process. We write this as

$$\alpha_{t+1} = T_t \alpha_t + W_t b + H_t u_t. \tag{9.10}$$

The transition equation is initialised at the first time point by setting

$$\alpha_1 = W_0 b + H_0 u_0. \tag{9.11}$$

The model is completed by making assumptions about the error process u_t and the vector of regressors b. We assume they are independently and normally distributed with

$$u_t \sim \mathsf{NID}(0, \sigma^2 I), \quad b = c + B\delta \quad \text{and} \quad \delta \sim \mathsf{N}(\mu, \sigma^2 \Lambda). \tag{9.12}$$

This formulation of the SSF is slightly different from that exploited in books such as Harvey (1989) and West and Harrison (1989). It is used to produce rather more elegant treatments of diffuse initial conditions and fixed effects; see de Jong (1991).

The state system matrices Z_t and T_t are fixed matrices which merely contain known values. The regression system matrices X_t and W_t are always known. The error system matrices G_t and H_t are also sparse but most non-zero values are unknown and are regarded as "hyper" parameters. The disturbance u_t is transformed into noise for the SSF by the error system matrices $\mathbf{G}_t$ and $\mathbf{H}_t$. All system matrices can be regarded as selection type matrices.

At first sight, the role for **b** in the SSF looks quite complicated. However, the SSF is written in this way to allow a unified treatment of a variety of features for time series models. There are three particularly important ones on which to focus. First, the measurement equation allows **b** to influence the observations directly through the regressors $\mathbf{X}_t$. In general $\mathbf{X}_t$, like $\mathbf{G}_t$ and $\mathbf{H}_t$, may be a sparse selection type matrix. Second, the states are directly affected by **b** through $\mathbf{W}_t$. This feature is particularly useful for handling interventions like structural changes (changes in trend) and growth changes (changes in slope). Third, the prior distribution of the initial state vector is (partly) defined via **b**.

The random vector **b** allows the analysis of models where diffuse distributions are placed on the regressors and on the non-stationary components of the initial state. The vector **b** is a linear combination of the known constant vector **c** and the random vector $\boldsymbol{\delta}$. The distribution of $\boldsymbol{\delta}$ is regarded as *diffuse*, i.e. the variance matrix Λ converges to infinity. Basically, this means that a proper prior distribution for the state cannot be given when components are non-stationary. For a technical discussion on these issues, we refer to de Jong and Chu-Chun-Lin (1994). The connection between diffuse initial conditions and marginal or restricted likelihoods is discussed in Tunnicliffe-Wilson (1989) and Shephard (1993). The unconditional distributions for stationary components, such as cycles, are handled via the second part of (9.11); that is, $\mathbf{H}_0 \mathbf{u}_0$.

The generality of the SSF can be simplified for structural time series models in three different ways.

(1) The constant vector **c** of the specification for **b** can be set to a zero vector such that $\mathbf{b} = \mathbf{B}\boldsymbol{\delta}$. The matrix **B** is a square selection matrix of zeros and ones. The vector $\boldsymbol{\delta}$ can be partitioned such that the regression effects and the initial effects are separated and that $\boldsymbol{\delta}$ enter the equations of the SSF directly. This is achieved by

$$\boldsymbol{\delta} = \begin{pmatrix} \boldsymbol{\delta}_x \\ \boldsymbol{\delta}_i \end{pmatrix}, \quad \text{and} \quad \mathbf{B} = (\mathbf{B}_x, \mathbf{B}_i),$$

where the $(k \times 1)$ vector $\boldsymbol{\delta}_x$ contains the regression effects, the $(d \times 1)$ vector $\boldsymbol{\delta}_i$ contains the initial effects and the dimensions of the matrices $\mathbf{B}$, $\mathbf{B}_x$ and $\mathbf{B}_i$ are $(k+d) \times (k+d)$, $(k+d) \times k$ and $(k+d) \times d$, respectively. This brings the specification for **b** to

$$\mathbf{b} = \mathbf{B}_x \boldsymbol{\delta}_x + \mathbf{B}_i \boldsymbol{\delta}_i$$

and

$$\begin{aligned}
\mathbf{X}_t \mathbf{b} &= \mathbf{X}_t^* \boldsymbol{\delta}_x & \text{where} \quad \mathbf{X}_t^* &= \mathbf{X}_t \mathbf{B}_x & \text{and} \quad \mathbf{X}_t \mathbf{B}_i &= 0, \\
\mathbf{W}_t \mathbf{b} &= \mathbf{W}_t^* \boldsymbol{\delta}_x & \text{where} \quad \mathbf{W}_t^* &= \mathbf{W}_t \mathbf{B}_x & \text{and} \quad \mathbf{W}_t \mathbf{B}_i &= 0, \\
\mathbf{W}_0 \mathbf{b} &= \mathbf{W}_0^* \boldsymbol{\delta}_i & \text{where} \quad \mathbf{W}_0^* &= \mathbf{W}_0 \mathbf{B}_i & \text{and} \quad \mathbf{W}_0 \mathbf{B}_x &= 0.
\end{aligned}$$

(2) For all models in STAMP, $\mathbf{G}_t$ and $\mathbf{H}_t$ will be orthogonal; that is, $\mathbf{G}'_t \mathbf{H}_t = 0$, so the noise terms in the two equations, $\mathbf{G}_t \mathbf{u}_t$ and $\mathbf{H}_t \mathbf{u}_t$, are independently distributed. This feature of a model simplifies the statistical analysis of the SSF.

(3) All models in STAMP can be handled by a time-invariant SSF. This implies that the t subscripts of the state and error system matrices can be dropped, i.e. $\mathbf{Z}_t = \mathbf{Z}$, $\mathbf{T}_t = \mathbf{T}$, $\mathbf{G}_t = \mathbf{G}$ and $\mathbf{H}_t = \mathbf{H}$. A time-invariant SSF also simplifies the statistical analysis of the SSF.

9.2.1 Structural time series models in SSF

Structural time series models fit nicely in the state space form. However, some care should be taken with the initial state vector specification.

9.2.1.1 Univariate models in SSF

Some examples are given to get a flavour of it.

- *Seasonal model.* The first example of a structural time series model in SSF is a model with smooth trend and trigonometric stochastic seasonals with $s = 4$ (quarterly observations). The SSF representation is

$$y_t = \begin{pmatrix} 1 & 0 & 1 & 0 & 0 \end{pmatrix} \alpha_t + \begin{pmatrix} \sigma_\epsilon & 0 & 0 & 0 & 0 \end{pmatrix} \mathbf{u}_t$$

$$\alpha_t = \begin{pmatrix} \mu_t \\ \beta_t \\ \gamma_{1,t} \\ \gamma_{2,t} \\ \gamma_{3,t} \end{pmatrix} = \begin{pmatrix} 1 & 1 & 0 & 0 & 0 \\ 0 & 1 & 0 & 0 & 0 \\ 0 & 0 & 0 & 1 & 0 \\ 0 & 0 & -1 & 0 & 0 \\ 0 & 0 & 0 & 0 & -1 \end{pmatrix} \alpha_{t-1} + \begin{pmatrix} 0 & 0 & 0 & 0 & 0 \\ 0 & \sigma_\zeta & 0 & 0 & 0 \\ 0 & 0 & \sigma_\omega & 0 & 0 \\ 0 & 0 & 0 & \sigma_\omega & 0 \\ 0 & 0 & 0 & 0 & \sigma_\omega \end{pmatrix} \mathbf{u}_t$$

- *Cycle model.* Another example is a univariate model which consists of a level and cycle component. The SSF representation of this model is

$$y_t = \begin{pmatrix} 1 & 1 & 0 \end{pmatrix} \alpha_t + \begin{pmatrix} \sigma_\epsilon & 0 & 0 & 0 \end{pmatrix} \mathbf{u}_t, \text{ where } \alpha_t = (\mu_t, \psi_t, \psi_t^*)',$$

$$\alpha_t = \begin{pmatrix} 1 & 0 & 0 \\ 0 & \rho \cos \lambda_c & \rho \sin \lambda_c \\ 0 & -\rho \sin \lambda_c & \rho \cos \lambda_c \end{pmatrix} \alpha_{t-1} + \begin{pmatrix} 0 & \sigma_\eta & 0 & 0 \\ 0 & 0 & \sigma_\kappa & 0 \\ 0 & 0 & 0 & \sigma_\kappa \end{pmatrix} \mathbf{u}_t$$

- *Explanatory variables.* A univariate structural time series model may include explanatory variables. Consider a local linear trend model with two explanatory variables and a level intervention at some time point $t = j$. The SSF representation of this model is given by

$$y_t = \begin{pmatrix} 1 & 0 \end{pmatrix} \alpha_t + \begin{pmatrix} x_{1,t} & x_{2,t} & 0 \end{pmatrix} \delta_x + \begin{pmatrix} \sigma_\epsilon & 0 & 0 \end{pmatrix} u_t$$

$$\alpha_t = \begin{pmatrix} \mu_t \\ \beta_t \end{pmatrix} = \begin{pmatrix} 1 & 1 \\ 0 & 1 \end{pmatrix} \alpha_{t-1} + \begin{pmatrix} 0 & 0 & w_t \\ 0 & 0 & 0 \end{pmatrix} \delta_x + \begin{pmatrix} 0 & \sigma_\eta & 0 \\ 0 & 0 & \sigma_\zeta \end{pmatrix} u_t$$

with vector $u_t = (\epsilon_t, \eta_t, \zeta_t)'$. The initial state vector is given by $\alpha_1 = \delta_i$ such that $B = I$ and $H_0 = 0$. The series $x_{1,t}$ and $x_{2,t}$ are explanatory variables and w_t is a series of zeros but at some time point $t = j$ its value is unity.

9.2.1.2 Multivariate model in SSF

Any multivariate model in STAMP with common factors can be written in the SUTSE form with lower rank variance matrices; see §9.1.4. The SUTSE form retains the sparse structure of the system matrices Z and T which is important for computational issues related to the estimation of these models. Therefore, this general formulation is adopted to specify a multivariate model in the SSF. Since the SUTSE form is a minor generalisation to the univariate model, only one example will be given. The multivariate model is a smooth trend plus cycle model which can be represented by the SSF as

$$y_t = \begin{pmatrix} I & 0 & I & 0 \end{pmatrix} \alpha_t + \begin{pmatrix} \Gamma_\epsilon & 0 & 0 & 0 \end{pmatrix} u_t,$$

$$\alpha_t = \begin{pmatrix} I & I & 0 & 0 \\ 0 & I & 0 & 0 \\ 0 & 0 & \rho \cos\lambda_c I & \rho \sin\lambda_c I \\ 0 & 0 & -\rho \sin\lambda_c I & \rho \cos\lambda_c I \end{pmatrix} \alpha_{t-1} + \begin{pmatrix} 0 & 0 & 0 & 0 \\ 0 & \Gamma_\zeta & 0 & 0 \\ 0 & 0 & \Gamma_\kappa & 0 \\ 0 & 0 & 0 & \Gamma_\kappa \end{pmatrix} u_t$$

where state vector $\alpha_t = (\mu'_t, \beta'_t, \psi'_t, \psi^{*'}_t)$, disturbance vector $u_t = (\epsilon_t, \zeta_t, \kappa_t, \kappa^*_t)'$ and each variance matrix is decomposed as $\Sigma = \Gamma\Gamma'$ with the appropriate subscript. The unknown values in transition matrix T rely on the *damping factor* ρ and the *frequency* λ_c. The initial state is given by

$$\alpha_1 = \begin{pmatrix} I & 0 \\ 0 & I \\ 0 & 0 \\ 0 & 0 \end{pmatrix} \delta_i + \frac{1}{\sqrt{1-\rho^2}} \begin{pmatrix} 0 & 0 \\ 0 & 0 \\ \Gamma_\kappa & 0 \\ 0 & \Gamma_\kappa \end{pmatrix} u_0$$

with vector $u_0 = (\kappa_0, \kappa^*_0)'$ and δ_i contains the initial effects for μ_t and β_t.

9.2.1.3 Deterministic trend and seasonal components

The deterministic seasonal component is the special case of the stochastic seasonal model with $\sigma_\omega = 0$. Alternatively, the fixed seasonal can be incorporated within $\mathbf{X_t}$ as it is usually done in regression models. The latter one is adopted in STAMP. In a similar way (partly) deterministic trends can be incorporated in $\mathbf{X_t}$ or $\mathbf{W_t}$. However, for practical reasons such as the implementation of trend interventions within $\mathbf{W_t}$, the trend components μ_t and β_t are always part of the state vector. The inclusion of a fixed trend in the state vector has some consequences for calculation and interpretation of residuals and the prediction error variance.

9.3 Kalman filter

The Kalman filter (KF) plays the same role for time series models in SSF as least squares computations for a regression model. The KF is primarily a set of vector and matrix recursions. The importance of the KF is based on

(1) computation of one-step ahead predictions of observation and state vectors, and the corresponding mean square errors;
(2) diagnostic checking by means of one-step ahead prediction errors;
(3) computation of the likelihood function via the one-step ahead prediction error decomposition;
(4) smoothing which uses the output of the KF.

The KF has a variety of forms, but the one exploited in STAMP computes

$$\mathbf{a}_{t|t-1} = E(\alpha_t|\mathbf{Y}_{t-1}, \boldsymbol{\delta}=\mathbf{0}),$$
$$\sigma^2 \mathbf{P}_{t|t-1} = E\{(\alpha_t - \mathbf{a}_{t|t-1})(\alpha_t - \mathbf{a}_{t|t-1})'|\mathbf{Y}_{t-1}, \boldsymbol{\delta}=\mathbf{0}\},$$

which are referred to as the mean and mean square error (MSE) of the state, respectively, given the past information and setting $\boldsymbol{\delta}$ to zero. Generally, we need to perform estimation with $\boldsymbol{\delta}$ not zero, so these terms will be corrected by the use of an augmented Kalman filter. This will be discussed in the next subsection.

The recursive equations of the Kalman filter are given by

$$\begin{aligned}
\mathbf{v}_t &= \mathbf{y}_t - \mathbf{X}_t \mathbf{b} - \mathbf{Z}\mathbf{a}_{t|t-1}, & \mathbf{F}_t &= \mathbf{Z}\mathbf{P}_{t|t-1}\mathbf{Z}' + \mathbf{GG}', \\
q_t &= q_{t-1} + \mathbf{v}_t' \mathbf{F}_t^{-1} \mathbf{v}_t, & \mathbf{K}_t &= \mathbf{T}\mathbf{P}_{t|t-1}\mathbf{Z}'\mathbf{F}_t^{-1}, \\
\mathbf{a}_{t+1|t} &= \mathbf{T}\mathbf{a}_{t|t-1} + \mathbf{W}_t \mathbf{b} + \mathbf{K}_t \mathbf{v}_t, \\
\mathbf{P}_{t+1|t} &= \mathbf{T}\mathbf{P}_{t|t-1}\mathbf{T}' - \mathbf{K}_t \mathbf{F}_t \mathbf{K}_t' + \mathbf{HH}',
\end{aligned}$$
(9.13)

starting with $\mathbf{a}_{1|0} = \mathbf{W}_0 \mathbf{b}$, $\mathbf{P}_{1|0} = \mathbf{H}_0 \mathbf{H}_0'$ and $q_0 = 0$. Here $\mathbf{K}_t$ is called the Kalman gain, while $\mathbf{v}_t$ and $\sigma^2 \mathbf{F}_t$ are the one-step ahead prediction error (or innovation) and its mean square error, respectively. The scaled innovations $\mathbf{F}_t^{-\frac{1}{2}} \mathbf{v}_t$ (or generalised least squares residuals) are approximately NID with zero mean and scale identity matrix as its variance matrix in a correctly specified model. Note that the estimate of σ^2 is given by $\widehat{\sigma}^2 = q_T/NT$. The proof of the KF is simple and it relies on standard results of linear estimation. Although the KF is presented in terms of time-invariant system and error matrices ($\mathbf{Z}$, $\mathbf{T}$, $\mathbf{G}$ and $\mathbf{H}$), the time indices can be added to these matrices without any problem.

In the special case of a univariate model, $N = 1$, the innovation vector and the corresponding mean square error matrix reduce to the scalars v_t and $\sigma^2 f_t$, respectively. Also, the Kalman gain matrix reduces to the vector $\mathbf{k_t}$. This notation is only used for univariate models.

The KF requires a positive (for univariate models) or a non-singular positive definite matrix (for multivariate models) $\mathbf{F_t}$ which may not occur at the start of the KF when, for example, the column rank of $\mathbf{G_1}$ is less N and $\mathbf{P_{1|0}} = \mathbf{H_0 H_0'} = \mathbf{0}$. An example is when no irregular is present in a structural time series model. In these circumstances, the KF is warmed up by a number of prediction updates; that is, $\mathbf{a_{j+1|0}} = \mathbf{Ta_{j|0}}$ and $\mathbf{P_{j+1|0}} = \mathbf{TP_{j|0}T'} + \mathbf{HH'}$. The initialisation process continues until $\mathbf{F_j} = \mathbf{ZP_j Z'} + \mathbf{GG'}$ is positive definite. The KF is then re-initialised by $\mathbf{a_{1|0}} = \mathbf{a_{i|0}}$, $\mathbf{P_{1|0}} = \mathbf{P_{i|0}}$ and $q_0 = 0$ where i is the number of required prediction updates.

When $\mathbf{F_t}$ becomes non-positive definite during the KF updating, it is generally assumed that a numerical problem has occurred. Available remedies like a *square-root* version of the KF are not considered. However, the implementation of the KF in STAMP ensures for each time-point t that the MSE matrices $\mathbf{P_{t+1|t}}$ and $\mathbf{F_t}$ are symmetric, which is crucial in avoiding numerical problems. Also, avoiding unnecessary computations, such as adding zeros and multiplying zeros or unity values, helps in this respect.

In case of a time-invariant SSF, the KF may reach a steady state solution at some time s; that is, when $\mathbf{P_{s+1|s}} = \mathbf{P_{s|s-1}}$ for $p < s \leq T$ where p is the dimension of the state vector. This also implies that $\mathbf{F_j}$ and $\mathbf{K_j}$ are time-invariant for $j = s, \ldots, T$. The KF reduces in these circumstances to the equations for $\mathbf{a_{t+1|t}}$, $\mathbf{v_t}$ and q_t which are computationally very cheap. Finally, the steady state can be interpreted as the solution of the *Riccati equation*; see Anderson and Moore (1979).

9.3.1 The augmented Kalman filter

The KF evaluates the one-step ahead prediction of the state vector conditional on $\delta = \mathbf{0}$. When non-stationary components and/or fixed regression effects are present in the model, the unknown random vector δ is treated as diffuse; see [14.2]. In these circumstances an augmented Kalman filter (AKF) is applied:

$$\begin{aligned} \mathbf{V_t} &= -\mathbf{ZA_{t|t-1}} - \mathbf{X_t B} \\ \mathbf{A_{t+1|t}} &= \mathbf{TA_{t|t-1}} + \mathbf{W_t B} + \mathbf{K_t V_t} \\ (\mathbf{s_t}, \mathbf{S_t}) &= (\mathbf{s_{t-1}}, \mathbf{S_{t-1}}) + \mathbf{V_t' F_t^{-1}}(\mathbf{v_t}, \mathbf{V_t}) \end{aligned} \qquad (9.14)$$

for $t = 1, \ldots, T$ and with $\mathbf{A}_{1|0} = \mathbf{W_0 B}$. The AKF must be regarded as a supplement to the KF equations; the expressions for $\mathbf{v_t}$, $\mathbf{F_t}$, $\mathbf{q_t}$, $\mathbf{K_t}$, $\mathbf{a}_{t+1|t}$ and $\mathbf{P}_{t+1|t}$ remain as they are. The number of columns for $\mathbf{V_t}$ and $\mathbf{A}_{t+1|t}$ is $k + d$, the same number of columns as in matrix $\mathbf{B}$ of the SSF. The one-step ahead prediction of the state vector and the corresponding MSE matrix are given by

$$\begin{aligned} E(\alpha_t | \mathbf{Y_{t-1}}) &= \widehat{\mathbf{a}}_{t|t-1} &= \mathbf{a}_{t|t-1} + \mathbf{A}_{t|t-1} \mathbf{S}_{t-1}^{-1} \mathbf{s}_{t-1}, \\ MSE(\alpha_t | \mathbf{Y_{t-1}}) &= \sigma^2 \widehat{\mathbf{P}}_{t|t-1} &= \sigma^2 (\mathbf{P}_{t|t-1} + \mathbf{A}_{t|t-1} \mathbf{S}_{t-1}^{-1} \mathbf{A}'_{t|t-1}), \end{aligned} \quad (9.15)$$

respectively. The one-step ahead prediction error and the corresponding MSE matrix (or variance matrix) are given by

$$\begin{aligned} \mathbf{y_t} - E(\mathbf{y_t} | \mathbf{Y_{t-1}}) &= \widehat{\mathbf{v}}_t &= \mathbf{v_t} + \mathbf{V_t} \mathbf{S}_{t-1}^{-1} \mathbf{s}_{t-1} \\ MSE(\mathbf{y_t} | \mathbf{Y_{t-1}}) &= \sigma^2 \widehat{\mathbf{F}}_t &= \sigma^2 (\mathbf{F_t} + \mathbf{V_t} \mathbf{S}_{t-1}^{-1} \mathbf{V}'_t), \end{aligned} \quad (9.16)$$

respectively. The matrix inversions for $\mathbf{S_t}$, $t = 1, \ldots, T$, can be evaluated recursively using similar methods as recursive regressions; see Appendix 9.7 of this chapter. The estimate of the scalar variance σ^2 is

$$\widehat{\sigma}^2 = \frac{1}{NT - d - k} \widehat{q}_T \quad \text{where} \quad \widehat{q}_T = q_T - \mathbf{s}'_\mathbf{T} \mathbf{S}_\mathbf{T}^{-1} \mathbf{s_T}. \quad (9.17)$$

The full sample generalised least squares estimate of δ and its MSE matrix are given by

$$\widehat{\delta} = \mathbf{S}_\mathbf{T}^{-1} \mathbf{s_T} \quad \text{and} \quad MSE(\widehat{\delta}) = \sigma^2 \mathbf{S}_\mathbf{T}^{-1} \quad (9.18)$$

respectively. The MSE matrix is used to obtain standard errors and t-statistics for $\widehat{\delta}$. In a similar way as vector δ, $\widehat{\delta}$ is partitioned as

$$\widehat{\delta} = \begin{pmatrix} \widehat{\delta}_x \\ \widehat{\delta}_i \end{pmatrix}.$$

For a proof and a full exposition of the augmented Kalman filter; see de Jong (1991). When the SSF does not contain values for $\mathbf{X_t}$ and $\mathbf{W_t}$, the KF can evaluate $\widehat{\mathbf{a}}_{t+1|t}$ from $t = d+1$ onwards. This implies that the AKF is only used for the first d updates. Then, the AKF is collapsed to the KF, mainly using equation (9.15) for $t = d$. On the other hand, when $\mathbf{X_t}$ and $\mathbf{W_t}$ are present in the SSF, a partial collapse with respect to δ_i is possible. The exact details of the full and the partial collapse are given in Appendix 9.7 of this chapter. Formal proofs on collapsing are given by de Jong and Chu-Chun-Lin (1994).

9.3.2 The likelihood function

The likelihood function can be obtained from the KF by using the innovations and their mean square errors. This is known as the *prediction error decomposition*. For example, the likelihood function, conditional on $\delta = 0$, is defined as

$$\log L(y|\theta^*, \delta = 0) = -\frac{NT}{2}\log \sigma^2 - \frac{1}{2}\sum_{t=1}^{T}\log|\mathbf{F_t}| - \frac{\mathbf{q_T}}{2\sigma^2} \qquad (9.19)$$

apart from some constant. To limit the number of logarithmic operations in computing the likelihood, it is important to check whether a steady-state has been reached such that $\log|\mathbf{F_t}|$ is constant for all t.

The unknown parameters in the system matrices of the SSF are stacked in vector θ. For a given vector $\theta = \theta^*$, the KF calculates the likelihood and, therefore, it provides means to obtain maximum likelihood estimates of the parameters. These matters are discussed in §9.6.

9.3.2.1 Univariate models: the concentrated likelihood

The diffuse likelihood function (DL) is defined as the likelihood when δ is supposed to be diffuse; see §9.2. For a given vector $\theta = \theta^*$, the DL is given by

$$\log L(y|\theta^*) = -\frac{T^*}{2}\log \sigma^2 - \frac{1}{2}\sum_{t=1}^{T}\log f_t - \frac{1}{2}\log|\mathbf{S_T}| - \frac{1}{2\sigma^2}\hat{\mathbf{q}}_T \qquad (9.20)$$

with $T^* = T - k - d$. When a (partial) collapse of the AKF has taken place at some time $t = m$, the DL is calculated by

$$\log L(y|\theta^*) = -\frac{T^*}{2}\log \sigma^2 - \frac{1}{2}\sum_{t=1}^{T}\log f_t - \frac{1}{2}\log|\mathbf{S_*}| - \frac{1}{2\sigma^2}\hat{\mathbf{q}}_T. \qquad (9.21)$$

Appendix 9.7 of this chapter defines $\mathbf{S_*}$ and it gives more details of likelihood calculation when a full or a partial collapse has taken place. The scalar variance σ^2 can be concentrated out of the likelihood DL. This concentrated diffuse likelihood function (CDL) is given by

$$\log L_c(y|\theta^*) = -\frac{T^*}{2}\log \hat{\sigma}^2 - \frac{1}{2}\sum_{t=1}^{T}\log f_t - \frac{1}{2}\log|\mathbf{S_*}|. \qquad (9.22)$$

To calculate the CDL, one variance corresponding to a particular unobserved component is set to a unity value such that this variance is set equal to σ^2 of (9.12). The maximum likelihood estimate of this 'concentrated out' variance is $\hat{\sigma}^2$. The other variances of the components in the model are now measured as a ratio to the concentrated variance. For example, in a local level model, where σ_ε^2 is concentrated out, we have $\sigma_\eta^2 = \sigma^2 q_\eta$. These variance ratios are referred to as *q-ratios*. Parameter estimation is carried out with respect to the *q-ratios*; see §9.6. The CDL is used for parameter estimation of univariate models in STAMP. For a full discussion on diffuse likelihoods, we refer to de Jong (1988).

9.3.2.2 Multivariate models

The diffuse likelihood is used for parameter estimation of multivariate models. Thus, no parameter is concentrated out from the likelihood. The AKF is used for calculation but it can be (partially) collapsed at some time $t = m$. The multivariate diffuse likelihood is defined as

$$\log L(y|\theta^*) = -\frac{1}{2}\sum_{t=1}^{T} \log |\mathbf{F_t}| - \frac{1}{2} \log |\mathbf{S}_*| - \frac{\widehat{q}_T}{2}, \qquad (9.23)$$

where $\mathbf{S}_*$ is defined in Appendix 9.7 of this chapter.

9.3.3 Prediction error variance

The prediction error variance (PEV) of a univariate structural time series model is the variance of the one-step ahead prediction errors in the steady state. As discussed at the start of this section, the steady state may require more updates of the KF than the number of available observations. The formal definition of the estimated PEV is

$$PEV = \widehat{\sigma}^2 f,$$

where f is the steady state value of f_t; that is, $f = \lim_{t \to \infty} f_t$. For a multivariate model, the prediction error variance matrix can be evaluated. When a steady state is not obtained, it may be useful to report a 'finite' PEV; that is,

$$PEV_n = \widehat{\sigma}^2 \widehat{f}_n,$$

where n is some large finite number. The 'finite' PEV at the end of the sample; that is, at $n = T$, might also be of interest.

9.3.4 The final state and regression estimates

The final state vector $\widehat{\mathbf{a}}_{T|T}$ contains, together with $\widehat{\boldsymbol{\delta}}_x$, all the information from the estimated SSF required to forecast future observations. Therefore, the final state is of special interest. The estimate of the state vector and its MSE matrix at the end of the sample are given by

$$\begin{aligned}
\widehat{\mathbf{a}}_{T|T} &= \widehat{\mathbf{a}}_{T|T-1} + \widehat{\mathbf{P}}_{T|T-1}\mathbf{Z}'_\mathbf{T}\widehat{\mathbf{F}}_\mathbf{T}^{-1}\widehat{\mathbf{v}}_\mathbf{T}, \\
\widehat{\sigma}^2\widehat{\mathbf{P}}_{T|T} &= \widehat{\sigma}^2(\widehat{\mathbf{P}}_{T|T-1} - \widehat{\mathbf{P}}_{T|T-1}\mathbf{Z}'_\mathbf{T}\widehat{\mathbf{F}}_\mathbf{T}^{-1}\mathbf{Z}_\mathbf{T}\widehat{\mathbf{P}}_{T|T-1}),
\end{aligned}$$

respectively. The diagonal part of the MSE matrix is used to form standard errors and t-statistics of the final state $\widehat{\mathbf{a}}_{T|T}$. Under certain circumstances, some elements of this diagonal part may be equal to zero and calculation of the t-statistic is not possible then. Also, for stationary parts of the state vector, it is not useful to calculate the t-statistic since stationary components are not persistent throughout the series.

The regression estimates are obtained from

$$\widehat{\boldsymbol{\delta}}_x = \overline{\mathbf{S}}_T^{-1} \overline{\mathbf{s}}_T \quad \text{and} \quad MSE(\widehat{\boldsymbol{\delta}}_x) = \widehat{\sigma}^2 \overline{\mathbf{S}}_T^{-1},$$

where $\overline{\mathbf{S}}_T$ and $\overline{\mathbf{s}}_T$ are defined in Appendix 9.7 of this chapter. The diagonal part of the MSE matrix is used to form standard errors and t-statistics.

9.3.5 Filtered components

The filtered components are based on the one-step ahead predictions of the state vector; see equation (9.15). When $k = 0$, the AKF is applied until $t = d$. Then, the AKF collapses to the KF which evaluates the filtered states directly. However, when $k > 0$, a partial collapse takes place and filtering is based on (9.15) with matrix $\overline{\mathbf{S}}_T^{-1}$ calculated recursively; see Appendix 9.7 of this chapter. Note that the filtered states are only defined for $t = d + k + 1, \ldots, T$.

9.3.6 Residuals

The residuals in STAMP for univariate models are defined as the standardised innovations

$$v_t = \widehat{v}_t / \widehat{\sigma} \widehat{f}_t^{\frac{1}{2}}, \quad t = d+1, \ldots, T. \tag{9.24}$$

where $\widehat{v}_t$ and $\widehat{\sigma}^2 \widehat{f}_t$ are defined as in (9.16). When $k = 0$, the residuals are obtained from the KF directly after the collapse of the AKF at $t = d$. In a multivariate model the residuals of the j-th equation are given by

$$v_{j,t} = \widehat{v}_{j,t} / \widehat{\sigma} \widehat{F}_{jj,t}^{\frac{1}{2}}$$

for $t = d + 1, \ldots, T$ where $a_{i,t}$ is the i-th element of vector $\mathbf{a_t}$ and $A_{ij,t}$ is the (i, j) element of matrix $\mathbf{A_t}$.

When a SSF contains fixed effects; that is, when $k > 0$, two sets of residuals are distinguished inside STAMP. These residuals are defined in the next two subsections. Note that the two sets are the same when $k = 0$.

9.3.6.1 Generalised least squares residuals

The generalised least squares (GLS) residuals are standardised innovations computed using (9.24), $t = d + 1, \ldots, T$, but where $\widehat{v}_t$ and $\widehat{f}_t$ are obtained from the AKF (9.15) applied to a SSF with b_x being replaced by $\mathbf{B_x}\widehat{\boldsymbol{\delta}}_x$. The calculation of these residuals requires two steps. Firstly, the estimates $\widehat{\boldsymbol{\delta}}_x$ and $\widehat{\sigma}$ are obtained from the AKF with $d+k$ columns for $\mathbf{A_{t+1|t}}$. Then, a second AKF is applied, with d columns for $\mathbf{A_{t+1|t}}$, from which $\widehat{\mathbf{v}}_t$ and $\widehat{\mathbf{f}}_t$ are obtained for $t = d + 1, \ldots, T$. Of course, the first AKF may be partially collapsed after $t = d$ and the second AKF may be fully collapsed to a KF after

$t = d$. The GLS residuals are not independently distributed with constant variance in small samples. Tests for normality, heteroskedasticity and serial correlation are applied to the GLS residuals.

9.3.6.2 Generalised recursive residuals

The generalised recursive residuals are the standardised residuals (9.24), but now obtained from an AKF, with $d + k$ columns for $\mathbf{A}_{t+1|t}$, accompanied with a set of recursive regressions for the inversion calculations of $\mathbf{S}_t$; see Appendix 9.7 of this chapter. A partial collapse of this AKF may take place at $t = d$. These residuals are denoted by w_t, for $t = d + k + 1, \ldots, T$, and they are commonly used in predictive tests and diagnostics.

9.4 Disturbance smoother

Smoothing refers to the estimation of the state vector $\boldsymbol{\alpha}_t$ and the disturbance vector $\mathbf{u}_t$ using information in the whole sample rather than just past data. Smoothing is an important feature because it is the basis for

(1) signal extraction, detrending and seasonal adjustment;
(2) diagnostic checking for detecting and distinguishing between outliers and structural changes using auxiliary residuals;
(3) EM algorithm for initial estimation of parameters;
(4) calculation of the score, defined as the first derivative of the likelihood with respect to the vector of parameters.

We consider the disturbance smoother (DS) of Koopman (1993) which directly estimates the disturbance vector of the SSF, $\mathbf{u}_t$, and the MSE matrix; that is,

$$E(\mathbf{u}_t|\mathbf{Y}_T, \boldsymbol{\delta} = 0) = \tilde{\mathbf{u}}_t \quad \text{and} \quad \text{MSE}(\tilde{\mathbf{u}}_t) = \sigma^2 \mathbf{C}_t,$$

respectively, with $t = 1, \ldots, T$. Note that in the context of structural time series models, the interest in $\mathbf{u}_t$ is limited to elements of $\mathbf{Gu}_t$ and $\mathbf{Hu}_t$ and their corresponding MSE matrices.

The DS is based on the same backwards recursions as found by Bryson and Ho (1969), de Jong (1989) and Kohn and Ansley (1989) which are given by

$$\begin{aligned} \mathbf{e}_t &= \mathbf{F}_t^{-1}\mathbf{v}_t - \mathbf{K}_t'\mathbf{r}_t, & \mathbf{r}_{t-1} &= \mathbf{Z}'\mathbf{e}_t + \mathbf{T}'\mathbf{r}_t, \\ \mathbf{D}_t &= \mathbf{F}_t^{-1} + \mathbf{K}_t'\mathbf{N}_t\mathbf{K}_t, & \mathbf{N}_{t-1} &= \mathbf{Z}_t'\mathbf{F}_t^{-1}\mathbf{Z}_t + \mathbf{L}_t'\mathbf{N}_t\mathbf{L}_t, \end{aligned}$$

where $\mathbf{L}_t = \mathbf{T} - \mathbf{K}_t\mathbf{Z}$ and $t = T, \ldots, 1$. These backwards recursions are initialised with $\mathbf{r}_T = \mathbf{0}$ and $\mathbf{N}_T = \mathbf{0}$. The storage requirements for the KF are limited to the quantities $\mathbf{v}_t$, $\mathbf{F}_t$ and $\mathbf{K}_t$ for $t = 1, \ldots, T$. The smooth estimates for the disturbances of the measurement equation and the MSE matrices are

$$\mathbf{G}\widetilde{\mathbf{u}}_t = \mathbf{GG}'\mathbf{e}_t \quad \text{and} \quad \sigma^2 \mathbf{GC}_t \mathbf{G}' = \sigma^2 (\mathbf{GG}' - \mathbf{GG}'\mathbf{D}_t \mathbf{GG}'), \qquad (9.25)$$

respectively. The smooth estimates for the disturbances of the transition equation and the mean square errors are

$$\mathbf{H}\widetilde{\mathbf{u}}_t = \mathbf{HH}'\mathbf{r}_t \quad \text{and} \quad \sigma^2 \mathbf{HC}_t \mathbf{H}' = \sigma^2 (\mathbf{HH}' - \mathbf{HH}'\mathbf{N}_t \mathbf{HH}'), \qquad (9.26)$$

respectively. The proof of the disturbance smoother is given by Koopman (1993). The DS is the most computationally efficient smoother available yet.

9.4.1 The augmented disturbance smoother

In a similar way as the KF, the DS must be augmented to allow for initial effects with diffuse distributions and fixed components. The supplement recursions for the DS are

$$\mathbf{E}_t = \mathbf{F}_t^{-1}\mathbf{V}_t - \mathbf{K}_t'\mathbf{R}_t \quad \text{and} \quad \mathbf{R}_{t-1} = \mathbf{Z}'\mathbf{E}_t + \mathbf{T}'\mathbf{R}_t,$$

and they will be referred to as the augmented disturbance smoother (ADS). The number of columns for $\mathbf{E}_t$ and $\mathbf{R}_t$ is equal to $d + k$. When the AKF is not collapsed at all, the full sample estimates of the disturbance vectors $E(\mathbf{Gu}_t|\mathbf{Y}_T)$, and the corresponding MSE matrices $MSE(\mathbf{Gu}_t|\mathbf{Y}_T)$, are given by (9.25) with $\mathbf{e}_t$ and $\mathbf{D}_t$ replaced by

$$\widehat{\mathbf{e}}_t = \mathbf{e}_t + \mathbf{E}_t \widetilde{\delta} \quad \text{and} \quad \widehat{\mathbf{D}}_t = \mathbf{D}_t - \mathbf{E}_t \mathbf{S}^{-1} \mathbf{E}_t', \qquad (9.27)$$

respectively, for $t = 1, \ldots, T$. In a similar way, $E(\mathbf{Hu}_t|\mathbf{Y}_T)$ and $MSE(\mathbf{Hu}_t|\mathbf{Y}_T)$ are given by (9.26) with $\mathbf{r}_t$ and $\mathbf{N}_t$ replaced by

$$\widehat{\mathbf{r}}_t = \mathbf{r}_t + \mathbf{R}_t \widetilde{\delta} \quad \text{and} \quad \widehat{\mathbf{N}}_t = \mathbf{N}_t - \mathbf{R}_t \mathbf{S}^{-1} \mathbf{R}_t', \qquad (9.28)$$

respectively. for the transition disturbances with Note that $\widetilde{\delta} = \mathbf{S}_T^{-1}\mathbf{s}_T$ and $\mathbf{S} = \mathbf{S}_T$. For a proof of the ADS, the reader is referred to Koopman (1993). When the AKF is fully collapsed to the KF at $t = m$, the DS can be used for $t = T, \ldots, m + 1$. The ADS should be applied for $t = m, \ldots, 1$ but when a collapse has taken place, the matrices $\mathbf{S}_T$ and $\mathbf{R}_m$ are not available. Therefore, smoothing is not straightforward when a collapse has occurred during the AKF. However, a suggestion of Chu-Chun-Lin and de Jong (1993) can be used to recover these missing matrices.

9.4.2 The EM algorithm and exact score

The EM algorithm is a recursive method to obtain maximum likelihood estimates for unknown elements in the state and error system matrices of the SSF; see Shumway and Stoffer (1982) and Watson and Engle (1983). This method requires a huge amount of computations and is not easy to implement. Koopman (1993) proposes a simple and computationally efficient EM algorithm for unknown values inside the variance matrices $\Omega_G = GG'$ and $\Omega_H = HH'$ where the error system matrices are supposed to be time-invariant. Let θ be the stack of these unknown values. Given a set of values for θ, say θ^*, the EM step calculates new variance matrices via

$$\Omega_G(\theta^\dagger) = \Omega_G(\theta^*) + \Omega_G(\theta^*)\Lambda_e\Omega_G(\theta^*),$$
$$\Omega_H(\theta^\dagger) = \Omega_H(\theta^*) + \Omega_H(\theta^*)\Lambda_r\Omega_H(\theta^*),$$

where

$$\Lambda_e = \sum_{t=1}^{T} \widehat{e}_t \widehat{e}_t' - \widehat{D}_t \quad \text{and} \quad \Lambda_r = \sum_{t=1}^{T} \widehat{r}_t \widehat{r}_t' - \widehat{N}_t.$$

The quantities $\widehat{e}_t$ and $\widehat{D}_t$ are defined in (9.27) and the quantities $\widehat{r}_t$ and $\widehat{N}_t$ are defined in (9.28). The log-likelihood value of the SSF with these new variance matrices is always larger compared to the log-likelihood value of the SSF with the old variance matrices. This EM algorithm only requires one pass through the KF and one pass back through the disturbance smoother. Note that the EM maximises the diffuse log-likelihood (9.21). For a proof and a discussion of this EM algorithm, the reader is referred to Koopman (1993).

The score vector is defined as the first derivative of the log-likelihood with respect to the vector of parameters θ. The vector of first derivatives is usually evaluated numerically. Koopman and Shephard (1992) present a method to calculate the exact score for any parameter in the state and error system matrices. However, the computational gain is only achieved for exact score calculations with respect to parameters in the error system matrices $G_t = G$ and $H_t = H$. Again, let θ be the stack of unknown values in the variance matrices of the SSF and let vector θ^* contain specific values for θ, then the exact score is given by

$$q(\theta^*) = \frac{\partial \log L(y|\theta^*)}{\partial \theta} = \frac{1}{2}\text{tr}\left\{\Lambda_e \frac{\partial \Omega_G(\theta^*)}{\partial \theta}\right\} + \frac{1}{2}\text{tr}\left\{\Lambda_r \frac{\partial \Omega_H(\theta^*)}{\partial \theta}\right\}, \quad (9.29)$$

and the definitions of the EM algorithm apply here as well.

9.4.3 Smoothed components

The smoothed state vector is defined as the estimated state vector using the full data-set; that is,

$$E(\alpha_t|Y_T) = \widetilde{\alpha}_t. \quad (9.30)$$

The smoothed state vector forms the basis for detrending and seasonal adjustment procedures for the SSF. Graphical inspection of a sub-set or a linear combination of the smoothed state vector is an important feature in STAMP. The actual values for (9.30) are obtained by

$$\widetilde{\alpha}_{t+1} = \mathbf{T}\widetilde{\alpha}_t + \mathbf{W}_t^* \widetilde{\delta}_x + \mathbf{H}\widetilde{u}_t, \qquad t = 1, \ldots, T, \qquad (9.31)$$

with $\widetilde{\alpha}_1 = \widehat{\mathbf{a}}_{1|0} + \widehat{\mathbf{P}}_{1|0}(\mathbf{r}_0 + \mathbf{R}_0 \widetilde{\delta})$. Hence, the smoothed state vector is calculated using the forward recursion (9.31) after the DS is applied and the vector $\mathbf{H}\widetilde{u}_t = \mathbf{HH}' \mathbf{r}_t$ is stored for $t = 1, \ldots, T$. The MSE matrix of the smoothed state vector is not considered; see de Jong (1989) and Kohn and Ansley (1989). It is shown by Koopman (1993) that this method of state smoothing is computationally very fast. Also note that no extra storage space is required since the storage space of the Kalman gain matrix $\mathbf{K}_t$ can be used to store the vector $\mathbf{H}\widetilde{u}_t$ for $t = 1, \ldots, T$.

9.4.4 Auxiliary residuals

The standardised smoothed estimates of the disturbances are referred to as *auxiliary residuals*. Graphs of these residuals, in conjunction with normality tests, are a valuable tool for detecting data irregularities such as outliers, level changes and slope changes. Note that the auxiliary residuals are serially correlated even when the model is correctly specified. Therefore, the normality test should be adjusted for the implied serial correlation. The technical details and some interesting applications can be found in Harvey and Koopman (1992).

The auxiliary residuals are obtained directly from (9.25) and (9.26). The residuals are standardised using the diagonal elements of the appropriate MSE matrix. The variance σ^2 must be replaced by its estimate $\widehat{\sigma}^2$.

9.5 Forecasting

An important feature of STAMP is its extensive forecast facility. The model-based extrapolations can be calculated for the actual series but also for any element of the state vector.

9.5.1 Forecasts of series and components

The forecast of the state vector l multi-steps ahead, for $l = 1, \ldots, L$, and its MSE matrix are given by

$$\widehat{\mathbf{a}}_{T+l|T} = E(\alpha_{T+l} | \mathbf{Y}_T), \qquad \sigma^2 \widehat{\mathbf{P}}_{T+l|T} = \text{MSE}(\alpha_{T+l} | \mathbf{Y}_T),$$

respectively. The forecasts and MSE matrices are generated recursively by

$$\begin{aligned} \widehat{\mathbf{a}}_{T+l|T} &= \mathbf{T}\widehat{\mathbf{a}}_{T+l-1|T} + \mathbf{W}_{T+l}^* \widetilde{\delta}_x, \\ \widehat{\sigma}^2 \widehat{\mathbf{P}}_{T+l|T} &= \widehat{\sigma}^2 (\mathbf{T}\widehat{\mathbf{P}}_{T+l-1|T} \mathbf{T}' + \mathbf{HH}'), \end{aligned} \qquad (9.32)$$

respectively. The initialisation is directly obtained from the AKF at time T. The forecast for $\mathbf{y}_{T+l}$ and the corresponding MSE matrices are given by

$$\begin{aligned}\widehat{\mathbf{y}}_{T+l|T} &= \mathbf{Z}\widehat{\mathbf{a}}_{T+l|T} + \mathbf{X}^*_{T+1}\widetilde{\boldsymbol{\delta}}_\mathbf{x}, \\ \widehat{\sigma}^2\widehat{\mathbf{F}}_{T+l|T} &= \widehat{\sigma}^2(\mathbf{Z}\widehat{\mathbf{P}}_{T+1|T}\mathbf{Z}' + \mathbf{GG}'),\end{aligned} \quad (9.33)$$

respectively. Graphical inspection of the forecasted values can be valuable for checking model validity.

9.5.2 Extrapolative residuals

The extrapolative residuals can be calculated when future values of $\mathbf{y}_t$ become available. They are defined as

$$\widehat{\mathbf{v}}_{t+l} = \mathbf{y}_{T+1} - \widehat{\mathbf{y}}_{T+1|T}, \quad (9.34)$$

where $\widehat{\mathbf{y}}_{T+l|T}$ is defined in (9.33). The variance matrices of the extrapolative residuals are equal to the MSE matrices of $\widehat{\mathbf{y}}_{T+l|T}$; that is, $\widehat{\sigma}^2\widehat{\mathbf{F}}_{T+l|T}$ as in (9.33). The extrapolative residuals can be standardised by using the corresponding diagonal elements of the variance matrix.

9.6 Parameter estimation

The model definitions are given in §9.1. It is shown in §9.2 how the models of STAMP can be put in SSF. The filtering and smoothing algorithms of §9.3 and §9.4, respectively, are associated with the SSF and can be applied conditional on known state and error system matrices. The unknown values inside these matrices are treated as parameters which need to be estimated. The main task of STAMP is to estimate these parameters using maximum likelihood estimation (MLE) methods. This subsection sets out the strategy for solving this problem.

9.6.1 The parameters of STAMP

The state and error system matrices of the SSF are time-invariant and contain many zero and unity values. Therefore, they can be regarded as selection matrices. For a univariate model, the unknown values in the error system matrices are the standard deviations of the unobserved components. The time-invariant transition matrix $\mathbf{T}$ of the SSF contains the unknown first-order autoregressive parameter of stationary components such as cycle and (optional) slope. In the case of a cycle, the state system matrix $\mathbf{T}$ also contains cosine and sine terms which rely on the unknown frequency λ. Table 14.2 overviews the set of possible parameters for a univariate model.

The variances are always positive, the cycle *damping factors* must be in the range $0 < \rho < 1$, and the *frequencies* can only take values in the range $0 < \lambda < \pi$. Since the parameters are to be estimated via a numerical method, some special transformations are used to enforce these restrictions on the parameters; see Table 14.2. The set of transformed parameters, relevant for a specific model, are stacked in the vector θ.

As observed from Table 14.2, the variance of the cycle, σ_ψ^2, is taken as the parameter instead of the variance of its disturbance σ_κ^2. In this case, $\sigma_\kappa^2 = \sigma_\psi^2(1 - \rho_\psi^2)^{\frac{1}{2}}$ and $\sigma_\kappa^2 \to 0$ as $\rho_\psi \to 1$. This allows the estimation of deterministic, but stationary, cycles which is the extreme case of $\rho_\psi = 1$.

Table 9.2 Parameters and transformations.

Variances	Parameter	Transformation
Irregular	σ_ϵ^2	$\exp(2\theta_\epsilon)$
Level	σ_η^2	$\exp(2\theta_\eta)$
Slope	σ_ζ^2	$\exp(2\theta_\zeta)$
Seasonal	σ_ω^2	$\exp(2\theta_\omega)$
Cycle	σ_ψ^2	$\exp(2\theta_\psi)$
AR(1)	σ_ξ^2	$\exp(2\theta_\xi)$

Autoregressive	Parameter	Transformation
Slope	ρ_β	$\|\theta_\beta\|(1+\theta_\beta^2)^{-\frac{1}{2}}$
Cycle	ρ_ψ	$\|\theta_\rho\|(1+\theta_\rho^2)^{-\frac{1}{2}}$
AR(1)	ρ_ν	$\theta_\nu(1+\theta_\nu^2)^{-\frac{1}{2}}$

Frequency	Parameter	Transformation
Cycle	λ	$2\pi/2 + \exp(\theta_\lambda)$

9.6.1.1 Variance matrices in a multivariate model

The same parameters exist for a multivariate model as for a univariate model, except that the variances are replaced by variance matrices; see §9.1. A special case is the vector autoregressive component which is discussed in §9.5.2. Since all variance matrices are treated the same, we consider a $N \times N$ matrix Σ, i.e. a non-negative symmetric matrix but not necessarily full rank. The *Cholesky decomposition* of Σ enforces these restrictions. The variance matrix is decomposed as $\Sigma = \Theta D \Theta'$ where $N \times p$ matrix Θ is a lower-triangular matrix with unity values on the leading diagonal, $1 \leq p \leq N$, and $p \times p$ matrix $\mathbf{D}$ is a non-negative diagonal matrix. The lower-triangular elements of Θ; that is, θ_{ij} with $i = 2, \ldots, N$ and $j = 1, \ldots, min(i-1, p)$, may take any value, including zero values. The diagonal elements of $\mathbf{D}$ are transformed as

$$d_i = \exp(2\theta_{ii})$$

where d_i is the i-th diagonal element of $\mathbf{D}$ and $i = 1, \ldots, p$. Indeed, matrix $\boldsymbol{\Theta}$ is the same *factor loadings* matrix as in the model specifications for common components; see §9.1.4. Since matrix $\boldsymbol{\Theta}$ has p columns, this would suggest that variance matrix $\boldsymbol{\Sigma}$ has rank p. In most cases this will be true. However, as any diagonal element of $\mathbf{D}$ can be zero, the rank of $\boldsymbol{\Sigma}$ can be less than p. Note that when $d_i = 0$, the i-th column of $\boldsymbol{\Theta}$ does not affect variance matrix $\boldsymbol{\Sigma}$ and can be set to zero.

The variance matrices are placed into the SSF via the error system matrices. For example, the variance matrix of the irregular enters the SSF via $\mathbf{G} = (\boldsymbol{\Theta}\mathbf{D}^{\frac{1}{2}}, \mathbf{0}, \ldots, \mathbf{0})$ where $\boldsymbol{\Theta}$ and $\mathbf{D}$ correspond to the decomposition of the irregular variance matrix $\boldsymbol{\Sigma}_\epsilon$ and $\mathbf{0}$ denotes a zero matrix.

9.6.1.2 Vector autoregressive components

The multivariate model may include a vector autoregression of order 1. The parameters inside the matrix $\boldsymbol{\Phi}$ appear in the SSF transition matrix $\mathbf{T}$. All roots of the matrix polynomial $\mathbf{I} - \boldsymbol{\Phi}\mathbf{L}$ must lie outside the unit circle. Within a numerical optimisation procedure, the constraints on $\boldsymbol{\Phi}$ needed to keep the process stationary are imposed by means of an algorithm given by Ansley and Kohn (1986). In the special case of a VAR(1) model, the transformations are:

- The parameters of the VAR component are put into the $N \times N$ unrestricted matrix $\mathbf{A}$ and the $N \times N$ positive definite symmetric matrix $\boldsymbol{\Gamma}$.
- Cholesky transformations are applied on the matrices $\mathbf{I} + \mathbf{A}\mathbf{A}' = \mathbf{B}\mathbf{B}'$ and $\boldsymbol{\Gamma} = \mathbf{L}\mathbf{L}'$.
- The VAR matrices are defined by $\boldsymbol{\Phi} = \mathbf{L}\mathbf{A}'\mathbf{B}'^{-1}\mathbf{L}^{-1}$ and $\boldsymbol{\Sigma}_v = \boldsymbol{\Gamma} - \boldsymbol{\Phi}\boldsymbol{\Gamma}\boldsymbol{\Phi}'$.

This method ensure a stationary VAR component. However, imposing restrictions on the VAR matrices may be difficult.

9.6.2 BFGS Estimation procedure*

The optimisation method is based on the BFGS *quasi-Newton* method and it aims to maximise the likelihood criterion $\ell(\theta)$. At every call to this updating scheme, it delivers a new estimate for θ, denoted by $\boldsymbol{\theta}_{i+1}$, given a current estimate, $\boldsymbol{\theta}_i$, which brings the object function $\ell(\theta)$ closer to its maximum. The updating scheme can be written as

$$\boldsymbol{\theta}_{i+1} = \boldsymbol{\theta}_i + s_i \boldsymbol{\delta}_i \qquad (9.35)$$

where s_i is a step size scalar and $\boldsymbol{\delta}_i$ is the search direction vector. Appendix 9.8 of this chapter gives some basic principles on numerical optimisation and in particular the BFGS updating scheme. Some practical details on initialisation, updating, linear search and termination, are given below.

- At the start of the estimation process an initial parameter vector $\boldsymbol{\theta_0}$ and an initial search direction $\boldsymbol{\delta_0}$ must be given. The procedure of setting the initial parameter vector is different for univariate models and for multivariate models; see §9.6.3 and §9.6.4, respectively. The initial direction vector is calculated as

$$\boldsymbol{\delta_0} = \mathbf{H}^{\mathrm{d}}(\boldsymbol{\theta_0})\mathbf{q}(\boldsymbol{\theta_0}) \qquad (9.36)$$

where $\mathbf{q}(\boldsymbol{\theta_0})$ is the score vector and $\mathbf{H}^{\mathrm{d}}(\boldsymbol{\theta_0})$ is the Hessian matrix which only contains values for the diagonal entries. These values are evaluated numerically and are based on parameter vector $\boldsymbol{\theta_0}$. The off-diagonal entries are zero. The initial score vector is evaluated numerically. For multivariate models, the score vector is evaluated analytically for any next update; see §9.4.2.

- The maximisation process is terminated when all three convergence criteria hold depending on a small value ε or when the number of BFGS updates is equal to the integer M or when irregularities occur. In the latter case, the estimation procedure is stopped with an error code attached to it. The three convergence criteria are

1. Likelihood: $crit_1 = |\ell(\boldsymbol{\theta_i}) - \ell(\boldsymbol{\theta_{i+1}})| / |\ell(\boldsymbol{\theta_i})| \quad < \quad \varepsilon$

2. Gradient: $crit_2 = \frac{1}{m}\sum_{j=1}^{m}|q_j(\boldsymbol{\theta_i})| \quad < \quad 10\varepsilon$

3. Parameter: $crit_3 = \frac{1}{m}\sum_{j=1}^{m}|\theta_{i+1,j} - \theta_{i,j}| / |\theta_{i,j}| \quad < \quad 100\varepsilon$

where $\theta_{i,j}$ denotes the j-th element of the parameter vector $\boldsymbol{\theta_i}$ and $q_j(\boldsymbol{\theta_i})$ denotes the j-th element of the score vector $\mathbf{q}(\boldsymbol{\theta_i})$. When all three convergence criteria hold, the estimation procedure terminates and returns one of the following messages

Very strong	$crit_1 < \varepsilon$	$crit_2 < \varepsilon$	$crit_3 < \varepsilon$
Strong	$crit_1 < \varepsilon$	$crit_2 < \varepsilon$	$crit_3 < 10^N \varepsilon$
Weak	$crit_1 < \varepsilon$	$crit_2 < 10^N \varepsilon$	$crit_3 < 10^N \varepsilon$
Very weak	$crit_1 < 10^N \varepsilon$	$crit_2 < 10^N \varepsilon$	$crit_3 < 10^N \varepsilon$

where N is the number of elements in the vector $\mathbf{y_t}$ of the state space form; see §9.2. For univariate models, N is equal to unity. The default values are $\varepsilon = 10^{-7}$ and $M = 100$.

- The direction vector $\boldsymbol{\delta_i}$ is obtained via the BFGS scheme; see Appendix 9.8 of this chapter. To ensure stability in the optimisation routine: the BFGS updating of (9.45) does not take place when $\mathbf{d'g}$ in (9.45) is smaller than ε.

- When the likelihood function, the score or the diagonal elements of the Hessian matrix cannot be evaluated for some reason, the estimation procedure terminates with an error message.

More specific details of implementation for univariate models and multivariate models are given below.

9.6.3 Estimation of univariate models*

For univariate models, STAMP maximises the concentrated diffuse log-likelihood $\log L_c(y|\theta)$ which is treated as an unconstrained non-linear function of θ, denoted by $\ell(\theta)$. The likelihood kernel $\ell(\theta)$ is appropriately scaled by a function of T, the number of observations.

9.6.3.1 Initial values

The elements of the vector of transformed parameters θ are given in Table 14.2. The elements related to standard deviations, θ^σ, are set equal to -0.5, except for θ_η (level) which initial value is -1 whereas θ_ζ (slope) and θ_ω (seasonal) are started off with -1.5 and -2, respectively.

One element of θ^σ needs to be concentrated out, which effectively means it must be restricted to zero (that is, a unity standard deviation). The choice of the concentrated parameter must be chosen appropriately. At the start of the optimisation routine, the element corresponding to the irregular standard deviation is concentrated out. When no irregular is present in the model we choose from level, slope, cycles, seasonal and autoregressive component, consecutively.

The initial values for ρ_ψ and λ, corresponding to a particular cycle, are provided by the user and STAMP transforms them automatically to θ_0. The initial values for θ_β and θ_υ are set to 2 (which corresponds to ρ values of approximately 0.9).

Before the quasi-Newton optimisation is started, a simple method is constructed to get θ^σ to the neighborhood of the optimal solution in a stable but quickly way. The initialisation method is similar to the quasi-Newton updating scheme (9.35). The j-th element of the direction vector $\boldsymbol{\delta_i}$ is given by

$$\delta_{i,j} = H_{j,j}(\boldsymbol{\theta_i})\mathbf{q_j}(\boldsymbol{\theta_i})$$

where $q_j(\boldsymbol{\theta_i})$ denotes the j-th element of the score vector $\mathbf{q}(\boldsymbol{\theta_i})$ and $H_{j,j}(\boldsymbol{\theta_i})$ denotes the (j,j) element of the Hessian matrix $\mathbf{H}(\boldsymbol{\theta_i})$. The score and the diagonal elements of the Hessian matrix are numerically evaluated at $\boldsymbol{\theta_i}$. The initialisation updating steps are repeated five times but it is stopped earlier if the stopping criterion

$$\sum_j \delta_{i,j} \nabla_j(\boldsymbol{\theta}_i) < 0.01$$

holds.

During this simple estimation procedure, a choice is made of which parameter should be concentrated out of the log-likelihood. At every step i, the parameter related to the largest value in vector $\boldsymbol{\theta_i}$, is concentrated out when its corresponding value in $\boldsymbol{\delta_i}$ is larger than 4. This rule reflects the desire to concentrate out the largest standard deviation of the likelihood. When the choice of concentrated parameter is changed, the other standard deviations should be adjusted, so they have the correct ratios with respect to the new concentrated parameter. This might create unstabilities in the search process. Therefore, the difference between the old and the new adjusted value is enforced to lie between -1 and 2. This strategy is a result of 'fine tuning' after it has been applied to a wide range of different models and data-sets.

9.6.3.2 Strategy for setting parameters fixed

During the estimation procedure, certain parameters might take values close to their boundaries. When a parameter moves closer and closer to its boundary, it may slow down the optimisation procedure as a whole. In these circumstances, it is advised to fix parameters at their boundary values but to allow estimation to be continued for the remaining parameters. The strategy for setting parameters fixed is as follows. In the case of standard deviations, when the transformed parameter value is smaller that -5 and its gradient value is smaller than 0.0001 (in absolute terms), it is set to a constant implying a zero standard deviation.

Similarly, the *damping factor* ρ of a cycle is set to unity, when the element corresponding to θ_ρ in vector $\boldsymbol{\theta_i}$ is larger than 25. If the value related to θ_λ (corresponding to the period of a cycle) is larger than 7 or less than -7, the program reduces the cycle component to an autoregressive component. In both cases, it is assumed that the corresponding gradient value is smaller than 0.0001 (in absolute terms). The transformed parameters related to ρ_β (slope) and ρ_v (autoregressive) don't apply to this strategy.

9.6.3.3 Strategy for determining the concentrated parameter

The initial estimation procedure has determined a concentrated standard deviation from θ^σ. However, during the main estimation process, it might be more appropriate to concentrate out another parameter. When the largest transformed parameter value (corresponding to θ^σ) exceeds the value 1.5, this particular parameter will be the new concentrated parameter. Then, the other parameters corresponding to θ^σ obtain values as ratios to the new concentrated parameter, the gradient vector $\mathbf{q}(\boldsymbol{\theta_i})$ is re-calculated and the Hessian matrix is replaced by the diagonal matrix $\mathbf{H^d}(\boldsymbol{\theta_i})$.

9.6.4 Estimation of multivariate models*

The estimation procedure for multivariate models follows mainly the same route as for univariate models but there are some differences. The likelihood kernel $\ell(\theta)$ is the diffuse log-likelihood $\log L(y|\theta)$ but properly scaled by a function of T. So there is no need to find an appropriate concentrated parameter. The EM algorithm plays a prominent role in the procedure to obtain initial variance matrices for the disturbances; see §9.4.2. The strategy of setting elements to zero in the Cholesky decompositions of the variance matrices is set out below. The parameters for the VAR component cannot be set fixed during the estimation procedure.

9.6.4.1 Initial values

The initial settings for the multivariate model are discussed in terms of the variance matrix, the Cholesky decomposition and the VAR coefficient matrix, rather than directly in terms of the parameter vector θ. The initialisation procedure automatically transforms the elements of these matrices into the vector θ_0.

The initial parameter settings of the variance matrices is similar to the univariate settings of the *q-ratios*, except that now a specific *q-ratio* applies to all N diagonal entries of the $\mathbf{D}$ matrix. The factor loading matrix Θ is set to the identity matrix. At this stage, the rank conditions of the variance matrices are ignored. The VAR matrix is set equal to a diagonal matrix with all entries equal to 0.9. The extra initial parameters for the cycle component are set manually by the user.

The Kalman filter is applied to the multivariate model and the sample variance matrix of the innovations is calculated. The variance matrices of the disturbances are set equal to this matrix but properly scaled by their appropriate *q-ratios*.

Then, the EM algorithm is applied; see §9.4.2. The number of EM steps is equal to 5 which has proved to be appropriate in most cases. The VAR matrix and the other stationary parameters (ρ and λ) remain fixed during the EM. After the EM steps the parameters are transformed as indicated in §9.6.1. For example, the Cholesky decomposition is used to get the diagonal matrix $\mathbf{D}$ and the factor loading matrix Θ from the variance matrix. At this stage, the rank conditions are taken into account. Finally, the parameters are stacked into θ_0.

9.6.4.2 Strategy for fixing parameters

The need to restrict parameters to their boundary values is important to speed up the process of estimation. This is even more relevant for multivariate models since the estimation process is likely to take much more time than univariate models because filtering and smoothing involve matrices with larger dimensions and more parameters need to be estimated.

The diagonal entries of $\mathbf{D}$ are treated in the same way as the standard deviations in a univariate model. The restrictions for the *damping factor* and the *frequency* parameter are also done in a similar fashion as for univariate models. No restrictions can be made with respect to the parameters related to the VAR component. The *factor loading* matrix elements enter the parameter vector without transformation.

When the j-th diagonal element of $\mathbf{D}$ is restricted to zero, the rank of the variance matrix Σ is reduced by one. All elements of the corresponding j-th column of the factor loading matrix can also be restricted to zero since these elements don't have any effect on the variance matrix Σ. Any specific element of the factor loading matrix Θ is set to zero when its value is smaller than 0.0001 and its corresponding gradient is also smaller than 0.0001 (both in absolute terms).

9.7 Appendix 1: Diffuse distributions*

This appendix discusses technical details about the way we treat initial and regression effects for the SSF. Most components in STAMP are initialised using diffuse priors and so we must treat these priors carefully in our calculations to ensure stable and exact results. This is carried out using the idea of augmented filtering and smoothing. To ensure computational efficiency the Kalman filter (KF) needs to be collapsed at some point.

9.7.1 Collapse of the augmented KF

For a SSF with $k = 0$; that is, $\mathbf{X_t} = \mathbf{0}$ and $\mathbf{W_t} = \mathbf{0}$, $t = 1, \ldots, T$, is treated by the AKF for $t = 1, \ldots, m$ and the AKF is collapsed to the KF using (9.15) at $t = m$. Then, the KF evaluates $\hat{\mathbf{a}}_{t+1|t}$ and $\hat{\mathbf{P}}_{t+1|t}$ directly, $t = 1, \ldots, T$. The collapse also contains the operation

$$\hat{q}_m = q_m - \mathbf{s}'_m \mathbf{S}_m^{-1} \mathbf{s}_m.$$

Indeed, the KF evaluates the residual vector $\hat{\mathbf{v}}_t$ and its (unscaled) variance matrix $\hat{\mathbf{F}}_t$ directly as well. The value for m is restricted by $T \geq m \geq d$ and $|\mathbf{S_m}| > 0$ but usually it is set to $m = d$.

For a SSF with $k > 0$, a full collapse cannot be made since $\mathbf{X_t}$ and $\mathbf{W_t}$ are time-varying. However, a partial collapse can take place with respect to δ_i. This means that the AKF is still applied but with a reduced number of columns. Therefore, we partition the AKF matrices

$$\begin{aligned} \mathbf{A}_{t|t-1} &= (\mathbf{A}_{\mathrm{x},t|t-1}, \; \mathbf{A}_{\mathrm{i},t|t-1}), & \mathbf{V_t} &= (\mathbf{V}_{\mathrm{x},t}, \; \mathbf{V}_{\mathrm{i},t}) \\ \mathbf{s_t} &= \begin{pmatrix} \mathbf{s}_{\mathrm{x},t} \\ \mathbf{s}_{\mathrm{i},t} \end{pmatrix}, & \mathbf{S_t} &= \begin{pmatrix} \mathbf{S}_{\mathrm{xx},t} & \mathbf{S}'_{\mathrm{ix},t} \\ \mathbf{S}_{\mathrm{ix},t} & \mathbf{S}_{\mathrm{ii},t} \end{pmatrix} \end{aligned} \quad (9.37)$$

When $\mathbf{S}_{\mathrm{ii},t}$ is invertible at $t = m$, the AKF can be partially collapsed to

$$\begin{aligned} \overline{\mathbf{A}}_{t+1|t} &= \mathbf{A}_{\mathrm{x},t+1|t} + \mathbf{A}_{\mathrm{i},t+1|t} \mathbf{S}_{\mathrm{ii},t}^{-1} \mathbf{S}_{\mathrm{ix},t}, \\ \overline{\mathbf{P}}_{t+1|t} &= \mathbf{P}_{t+1|t} + \mathbf{A}_{\mathrm{i},t+1|t} \mathbf{S}_{\mathrm{ii},t}^{-1} \mathbf{A}'_{\mathrm{i},t+1|t}, \end{aligned}$$

and

$$\begin{aligned} \overline{q}_t &= q_t - \mathbf{s}'_{\mathrm{i},t} \mathbf{S}_{\mathrm{ii},t}^{-1} \mathbf{s}_{\mathrm{i},t}, \\ \overline{\mathbf{s}}_t &= \mathbf{s}_{\mathrm{x},t} - \mathbf{S}'_{\mathrm{ix},t} \mathbf{S}_{\mathrm{ii},t}^{-1} \mathbf{s}_{\mathrm{i},t}, \\ \overline{\mathbf{S}}_t &= \mathbf{S}_{\mathrm{xx},t} - \mathbf{S}'_{\mathrm{ix},t} \mathbf{S}_{\mathrm{ii},t}^{-1} \mathbf{S}_{\mathrm{ix},t}. \end{aligned}$$

Note that matrix $\mathbf{B}$ of the (partially collapsed) AKF is replaced by $\overline{\mathbf{B}} = \mathbf{B_x}$. The matrices $\overline{\mathbf{A}}_{t+1|t}, \overline{\mathbf{V}}_t = -\mathbf{Z}_t \overline{\mathbf{A}}_{t+1|t} - \mathbf{X}_t \overline{\mathbf{B}}$ and $\overline{\mathbf{B}}$ have k columns.

9.7.2 Likelihood calculation

At the end of a fully collapsed AKF, the estimate of σ^2 is given by $\hat{\sigma}^2 = \hat{q}_T/(NT - d - k)$ where $k = 0$ and $\hat{q}_T$ is obtained from the collapsed KF. When $k > 0$, $\hat{q}_T$ is given by

$$\hat{q}_T = \overline{q}_T - \overline{\mathbf{s}}_\mathbf{T}' \overline{\mathbf{S}}_\mathbf{T}^{-1} \overline{\mathbf{s}}_\mathbf{T}.$$

Likelihood calculation also requires the log value of the term $|\mathbf{S}_*|$ where $\mathbf{S}_* = \mathbf{S_T}$. If $k = 0$ and a full collapse has taken place at $t = m$, then $|\mathbf{S}_*| = |\mathbf{S_m}|$. If $k > 0$ and a partial collapse has taken place at $t = m$, then $|\mathbf{S}_*| = |\mathbf{S_{ii,m}}||\overline{\mathbf{S}}_\mathbf{T}|$. Note that the determinants are obtained without any extra computational costs since the inverses of these matrices are calculated earlier.

9.7.3 Recursive regressions

The recursive evaluation of least squares estimates is common practice in standard regression analysis since it is a way to get the recursive residuals. The same technique can be applied to the recursive evaluation of $\mathbf{S_t}$ which consists of the equations

$$\begin{aligned}
\hat{\mathbf{a}}_{t|t-1} &= \mathbf{a}_{t|t-1} + \mathbf{A}_{t|t-1}\mathbf{c}_{t-1}, & \hat{\mathbf{P}}_{t|t-1} &= \mathbf{P}_{t|t-1} + \mathbf{A}_{t|t-1}\mathbf{C}_{t-1}\mathbf{A}'_{t|t-1}, \\
\hat{\mathbf{v}}_t &= \mathbf{v}_t + \mathbf{V}_t \mathbf{c}_{t-1}, & \hat{\mathbf{F}}_t &= \mathbf{F}_t + \mathbf{V}_t \mathbf{C}_{t-1}\mathbf{V}'_t, \\
& & \mathbf{K}^*_t &= \hat{\mathbf{F}}_t^{-1} \mathbf{V}_t \mathbf{C}_{t-1}, \\
\mathbf{c}_t &= \mathbf{c}_{t-1} + \mathbf{K}^{*\prime}_t \hat{\mathbf{v}}_t, & \mathbf{C}_t &= \mathbf{C}_{t-1} - \mathbf{K}'_t \hat{\mathbf{F}}_t \mathbf{K}^*_t.
\end{aligned}$$
(9.38)

where $\mathbf{c_t} = \mathbf{S}_t^{-1}\mathbf{s}_t$ and $\mathbf{C_t} = \mathbf{S}_t^{-1}$.

9.8 Appendix 2: Numerical optimisation*

Numerical optimisation methods are used to obtain the maximum likelihood estimates for the parameter vector θ. Here we outline some of the basic principles behind the estimation procedure of STAMP.

There is a vast literature on non-linear optimisation techniques (see, among many others, Gill, Murray and Wright (1981), Cramer (1986) and Thisted (1988)). Note that many texts on optimisation focus on minimisation, rather than maximisation, but of course: $\max \ell(\theta) = -\min -\ell(\theta)$.

A first approach to obtaining the MLE θ from $\ell(\theta)$ is to consider solving the score equations, assuming the relevant partial derivatives exist:

$$\nabla \ell(\theta) = \frac{\partial \ell(\theta)}{\partial \theta} = \mathbf{q}(\theta). \tag{9.39}$$

Then $\mathbf{q}(\theta) = 0$ defines the necessary conditions for a local maximum of $\ell(\theta)$ at θ. A sufficient condition is that

$$\nabla^2 \ell(\theta) = \frac{\partial^2 \ell(\theta)}{\partial \theta \partial \theta'} = \frac{\partial \mathbf{q}(\theta)'}{\partial \theta} = -\mathbf{Q}(\theta) \qquad (9.40)$$

also exists and is negative definite at θ. If $\mathbf{Q}(\cdot)$ is positive definite for all parameter values, the likelihood is concave, and hence has a unique maximum; if not, there could be local optima or singularities.

9.8.1 Newton type methods

When $\mathbf{q}(\theta)$ is a set of equations that are linear in θ, $\mathbf{q}(\theta) = \mathbf{0}$ can be solved explicitly for θ. More generally, $\mathbf{q}(\theta)$ is non-linear, yielding a problem of locating θ which is no more tractable than maximising $\ell(\theta)$. Thus, we consider iterative approaches in which a sequence of values of θ (denoted θ_i at the ith iteration) is obtained approximating $\mathbf{q}(\theta_i) = \mathbf{0}$, and corresponding to non-decreasing values of the criterion function $\ell(\cdot)$: $\ell(\theta_{i+1}) \geq \ell(\theta_i)$:

$$\theta_{i+1} = \mathbf{h}(\theta_i) \text{ for } i = 1, 2, \ldots, I \leq M \qquad (9.41)$$

where M is a terminal maximum number of steps, from an initial value θ_0. A convergence criterion is used to terminate the iteration, such as $\mathbf{q}(\theta_{i+1}) \simeq \mathbf{0}$ or $|\ell(\theta_{i+1}) - \ell(\theta_i)| \leq \epsilon$. These implementation-specific aspects are discussed in §9.6.

Expand $\mathbf{q}(\theta) = \mathbf{0}$ in a first-order Taylor's series:

$$\mathbf{q}(\theta) \simeq \mathbf{q}(\theta_1) - \mathbf{Q}(\theta_1)(\theta - \theta_1) = \mathbf{0}, \qquad (9.42)$$

written as the iterative rule:

$$\theta_{i+1} = \theta_i + \mathbf{H}(\theta_i)\mathbf{q}(\theta_i), \qquad (9.43)$$

where $\mathbf{H}(\cdot) = \mathbf{Q}^{-1}(\cdot)$ is the *Hessian* matrix. The gradient, $\mathbf{q}(\theta_i)$, determines the direction of the step to be taken, and $\mathbf{H}(\theta_i)$ modifies the size of the step which determines the metric: this algorithm is the Newton-Raphson technique, or *Newton's method*. Even when the direction is uphill, it is possible to overstep the maximum in that direction. In that case it is essential to add a line search to determine a step length $s_i \in [0, 1]$:

$$\theta_{i+1} = \theta_i + s_i \delta_i, \qquad (9.44)$$

where $\delta_i = \mathbf{H}(\theta_i)\mathbf{q}(\theta_i)$ and s_i is chosen to ensure that $\ell(\theta_{i+1}) \geq \ell(\theta_i)$.

STAMP uses the so-called variable metric or Broyden-Fletcher-Goldfarb-Shanno (BFGS) update *quasi-Newton* methods. These approximate the *Hessian* matrix $\mathbf{H}(\theta_i)$ by a symmetric positive definite matrix $\mathbf{K}_i$ which is updated at every iteration but converges on $\mathbf{H}(\cdot)$.

Let $\mathbf{d} = s_i \delta_i = \theta_i - \theta_{i-1}$ and $\mathbf{g} = \mathbf{q}(\theta_i) - \mathbf{q}(\theta_{i-1})$; then the BFGS update is:

$$\mathbf{K}_{i+1} = \mathbf{K}_i + \left(1 + \frac{\mathbf{g}'\mathbf{K}_i\mathbf{g}}{\mathbf{d}'\mathbf{g}}\right)\frac{\mathbf{dd}'}{\mathbf{d}'\mathbf{g}} - \frac{\mathbf{dg}'\mathbf{K}_i + \mathbf{K}_i\mathbf{gd}'}{\mathbf{d}'\mathbf{g}}. \qquad (9.45)$$

This satisfies the quasi-Newton condition $\mathbf{K}_{i+1}\mathbf{g} = \mathbf{d}$, and possesses the properties of hereditary symmetry ($\mathbf{K}_{i+1}$ is symmetric if $\mathbf{K}_i$ is), hereditary positive definiteness, and super-linear convergence.

9.8.2 Numerical score vector and diagonal Hessian matrix

When the analytic formula for $\mathbf{q}(\boldsymbol{\theta})$ cannot be obtained easily, STAMP uses BFGS with a numerical approximation to $\mathbf{q}(\cdot)$ based on finite difference approximations. The numerical derivatives are calculated using:

$$q_j(\boldsymbol{\theta}) = \frac{\partial \ell(\boldsymbol{\theta})}{\partial \left(\iota_j' \boldsymbol{\theta}\right)} \simeq \frac{\ell(\boldsymbol{\theta} + \varepsilon \iota_j) - \ell(\boldsymbol{\theta})}{\varepsilon} \tag{9.46}$$

where ι_j is a unit vector (for example, $(0\ 1\ 0\ \ldots\ 0)'$ for $j = 2$), ε is a suitably chosen step length. Thus, ε represents a compromise between round-off error (cancellation of leading digits when subtracting nearly equal numbers) and truncation error (ignoring terms of higher order than ε in the approximation).

The diagonal entries of the Hessian matrix are numerically evaluated as

$$H_{j,j}^d(\boldsymbol{\theta}) = -\frac{1}{Q_{j,j}(\boldsymbol{\theta})}$$

where

$$Q_{j,j}(\boldsymbol{\theta}) = \frac{\partial^2 \ell(\boldsymbol{\theta})}{\partial \left(\iota_j' \boldsymbol{\theta}\right) \partial \left(\iota_j' \boldsymbol{\theta}\right)} = \frac{\partial q_j(\boldsymbol{\theta})}{\partial \left(\iota_j' \boldsymbol{\theta}\right)} \simeq \frac{q_j(\boldsymbol{\theta} + \varepsilon \iota_j) - q_j(\boldsymbol{\theta})}{\varepsilon}$$

with the same step-length ε.

Chapter 10

Statistical Model Output

This chapter defines the output of STAMP. The first section explains when and where the various parts of the output appear. The remaining sections explain the output is presented for every option of the **Test** menu.

10.1 Output from STAMP

This section details the output which occurs when a model is fitted to data. There are three different forms of output. Firstly, the iterations of the numerical optimisation algorithm are displayed in the Message window. Secondly, once the estimation algorithm has stopped (hopefully having converged!), the Results window reports details of the model specification and the results of the optimisation algorithm, and gives a summary of diagnostic statistics. Finally, the **Test** menu can be used to prompt STAMP to output specific details of the parameter estimates, state variables, model fit and diagnostics.

10.1.1 Model estimation

In order to compute the maximum likelihood estimates, STAMP first computes initial estimates and then it moves on to maximise the likelihood function directly by numerical optimisation; see Chapter 9. During this second phase various statistics are reported in the Message window:

- *Parameters* - the current values of the transformed parameters;
- *Gradients* - the derivatives of the likelihood function with respect to the transformed parameters;
- *Function value* - the kernel of the log-likelihood function;
- *Initial value* - the value of the function after the initial phase;
- *% change* - the percentage increase of the current likelihood kernel over the initial value.

10.1.2 Selected model and estimation output

The initial output, written to the Results window, gives details of the selected model and various results from the numerical optimisation procedure; an example is reproduced in §8.3.1. In particular, the selected components are listed and the number of parameters is given. The sample period used for parameter estimation is also given.

A variety of statistics emerge from the numerical optimisation. The likelihood kernel evaluated at $\hat{\theta}$ is reported together with the number of iterations required for convergence. Three different convergence measures are reported and these are categorised by some term like STRONG CONVERGENCE; see §9.6.

10.1.3 Summary statistics

An example of the diagnostic summary report, which is printed to the Results window after fitting, is given in §8.3.2. Broadly this summary gives the basic diagnostics and goodness-of-fit statistics. The precise definitions are given in §10.4 and §10.7.

- *Log-likelihood* - this is the actual log-likelihood function at its maximum;
- *Prediction error variance (PEV)* - the variance, or variance matrix, of the one-step ahead prediction errors in the steady state, or the finite PEV if accompanied by a message of 'No steady state';
- *Std. Error* - the equation standard error is the square root of the PEV;
- *Normality* - the Doornik–Hansen statistic, which is the Bowman–Shenton statistic with the correction of Doornik and Hansen (1994), distributed approximately as χ^2_2 under the null hypothesis;
- $H(h)$ - a test for heteroskedasticity, distributed approximately as $F(h, h)$;
- $r(\tau)$ - the residual autocorrelation at lag τ, distributed approximately as $N(0, 1/T)$;
- DW - Durbin-Watson statistic, distributed approximately as $N(2, 4/T)$;
- $Q(P, d)$ - Box–Ljung Q-statistic based on the first P residual autocorrelations and distributed approximately as χ^2_d;
- R^2, R^2_D or R^2_S - the most suitable coefficient of determination.

Notice that the loss in the degrees of freedom d of the Box–Ljung Q-statistic takes account of the number of relative parameters in θ. It does not take account of any lagged dependent variables and their presence necessitates further adjustment.

10.1.4 The sample period

The parameters are estimated using the *model sample* as recorded from the Estimation dialog in the **Model** menu. The calculation of diagnostics and goodness-of-fit measures are calculated using the *state sample*. The default is that the model sample and the state sample are the same. However, the state sample might be changed to check subsamples isolated from the rest of the sample. This can be done in the option Final state. Of course, both samples must be within the boundaries of the *data sample*.

10.1.5 Test

The nine items of the **Test** menu are self-explanatory. The headings of the following sections correspond to them. Note that the output for Goodness of Fit is requested by the Final state option and that many of the conventional time series diagnostics (built out of the innovations, or one-step ahead prediction errors) will come up in the Residual option.

10.2 Parameters

The variance parameters together with frequencies and damping factors for cycles and coefficients for autoregressive components are referred to as parameters. The maximum likelihood estimates of the parameters can be requested from the More written output dialog. This dialog also possibly outputs the transformed parameters with their scores and approximate standard errors.

10.2.1 Variances and standard deviations

This gives the variances of the disturbances driving the various components. Thus, the level variance is σ_η^2, the slope variance is σ_ζ^2 and so on. The standard deviations are the square roots of the variances. The *q-ratios* are the ratios of each variance or standard deviation to the largest.

10.2.2 Cycle and AR(1)

For a stochastic cycle component, this option outputs the damping factor ρ, the frequency λ, the period $2\pi/\lambda$, and the period in terms of 'years'; that is, $2\pi/s\lambda$, where s is the number of 'seasons'. In addition it gives the variance of the cycle, $\sigma_\psi^2 = \sigma_\kappa^2/(1 - \rho^2)$. For the AR(1) component in a univariate model, the autoregressive coefficient is reported.

10.2.3 Transformed parameters and standard errors

The vector of estimated parameters, which are transformed as in §9.6, can be written to the Results window using this option. It also reports the score vector, associated with the estimated vector of parameters, and the numerical standard errors of the transformed parameters.

The scores are calculated as described in §9.6. The standard errors are obtained from the diagonal entries of the numerically evaluated Hessian matrix **Q**. Formally, the standard error of the i-th estimated parameter is defined as

$$se_i = -1/\sqrt{Q_i}$$

where Q_i is the i-th diagonal entry of matrix **Q**.

10.3 Final state

The estimate of the final state vector contains all the latest information on the components in the model. Together with the regression estimates of the explanatory variables and the interventions, the final state contains all the information needed to make forecasts.

The Final state dialog also provides useful information about the estimated seasonal pattern and the amplitude of the cycle. Some goodness-of-fit statistics can be generated, but these are discussed in §10.4.

10.3.1 Analysis of state

Following the notation as in Koopman, Shephard and Doornik (1999), the final state is the filtered estimate at time T; that is, $\hat{\mathbf{a}}_{T|T}$. The square roots of the diagonal elements of the corresponding MSE matrix, $\hat{\sigma}^2 \hat{\mathbf{P}}_{T|T}$, are the root mean square errors (RMSEs) of the corresponding state elements. The *t-value* is the estimate of the state divided by the RMSE. Applied to the i-th element of the state, this is

$$t\text{-}value_i = a_i \,/\, RMSE_i.$$

where a_i is the i-th element of vector $\hat{\mathbf{a}}_{T|T}$.

The Prob.values shows the probability of the absolute value of a standard normal variable exceeding the absolute value of the *t-value*. In a hypothesis testing situation, a Prob.value of less than 0.05 corresponds to rejection of the hypothesis of a zero value on a two-sided test at the 5% level of significance. A Prob.value of less than 0.05 is indicated by *, while a value less than 0.01 is denoted **.

The various unobserved components which make up the structural time series model all appear in the final state .

- *Trend* - 'Lvl' is the estimate of the level of the trend μ_T while 'Slp' is the estimate of the slope β_T. The *t-value* might not appear for the level when no irregular is specified in the model or when the irregular variance is estimated as zero.
- *Trigonometric seasonal* - the final state contains the estimates for $\gamma_{1,T}$, $\gamma_{1,T}^*$, $\gamma_{2,T}$, $\gamma_{2,T}^*$, etc. and are denoted in the output by 'Sea_1', 'Sea_2' and so on. Note

that the meaning of 'Sea_1', 'Sea_2', etc., is different to what it is in dummy variable seasonality, and so the final estimated seasonal pattern cannot be identified directly. It must be requested in 'Seasonal tests'.
- *Cycle* - the terms 'Cyk_1' and 'Cyk_2' give the estimates of ψ_T and ψ_T^* for cycle k, where k is 1, 2 and/or 3. Since a cycle component is not persistent throughout the series, a *t-value* is not appropriate here.
- *Autoregressive* - the estimate of v_T is denoted as 'Ar1'.

10.3.2 Regression analysis

The estimates of the regression parameters, the parameters of the intervention variables and the fixed seasonal dummies are computed as described in Koopman, Shephard and Doornik (1999). These estimates are reported upon request (if part of the specified model) and they may be interpreted in much the same way as in a standard regression model. Because they are deterministic (not time-varying), their RMSEs are standard deviations. The *t-values* would have t-distributions if the (relative) parameters were known. Although they are asymptotically normal, a t-distribution may provide a better approximation to small sample properties. As was noted earlier, the Prob. values are based on the normal distribution.

10.3.3 Seasonal tests

If the seasonal component is deterministic, either because it is specified to be 'fixed' at the outset or its disturbance variance is estimated to be zero, a joint test of significance can be carried out on the $s - 1$ seasonal effects. The test is essentially the same as a test for the joint significance of a set of explanatory variables in regression. Under the null hypothesis of no seasonality, the large sample distribution of the test statistic, denoted by 'Seasonal Chi$^2(s - 1)$' in the output, is χ^2_{s-1}. The Prob. value is the probability of a χ^2_{s-1} variable exceeding the value of the test statistic. In the case of a stochastic seasonal, the joint seasonal test is also produced although a formal joint test of significance of the seasonal effect is inappropriate. However, the seasonal pattern is persistent throughout the series and when the seasonal pattern changes relatively slowly, which is usually the case, the test statistic can provide a useful guide to the relative importance of the seasonal.

The formal definition of the test statistic is

$$\mathbf{a}' \mathbf{P}^{-1} \mathbf{a}$$

where **a** contains the estimates of the $s - 1$ seasonal effects at time T and **P** is the corresponding MSE matrix. In the case of a fixed seasonal, **a** and **P** are obtained from the matrix $(\mathbf{s_T}, \widehat{\sigma}^2 \mathbf{S_T})$. For a stochastic seasonal, **a** and **P** come from the matrix $(\widehat{\mathbf{a}}_{T|T}, \widehat{\sigma}^2 \widehat{\mathbf{P}}_{T|T})$; see Koopman, Shephard and Doornik (1999).

The final estimate of the seasonal effect for each season; that is, the estimated seasonal pattern at time T, is given here as well.

10.3.4 Cycle tests

The *amplitude* of the cycle is

$$\sqrt{a^2 + a^{*2}}$$

where a is the element of $\hat{\mathbf{a}}_{T|T}$ corresponding to ψ_T and a^* corresponds to ψ_T^*. The reported ratio is the comparison of the *amplitude* of the cycle with the level of the trend, and is also reported. This gives an indication of its relative importance. When the cycle is deterministic, but stationary, a joint significance χ^2 test (the same as the seasonal test) is produced as well.

10.3.5 Data in logs

When it is indicated that a model has been estimated in logarithms, some additional output for specific components appears in the Results window.

- *Trend* - The value $\exp(a)$ is reported where a is the element of $\hat{\mathbf{a}}_{T|T}$ corresponding to μ_T. When a slope component is specified, the estimated annual growth rate is reported as well. The annual growth rate is the estimate of β_T multiplied by $100s$.
- *Seasonal* - The multiplicative effect of the seasonal pattern on the level of the series is given by $\exp(a_j)$, where a_j is the estimate of the seasonal effect in season j. Also the percentage effect is calculated; that is, $100\{\exp(a_j) - 1\}$. This output is added to the output generated for the option 'Seasonal tests'.
- *Cycle* - Additional to Cycle tests, the amplitude as a percentage of the level is computed. This is simply multiplying the estimated amplitude by 100.

10.4 Goodness of fit

The goodness-of-fit statistics in STAMP all relate to the residuals, which are the standardised innovations; see Koopman, Shephard and Doornik (1999). When explanatory variables are specified in the model, the generalised least squares residuals are used. The diagnostic checking facility for these residuals are provided in the Residuals option; see §10.7.

10.4.1 Prediction error variance

The basic measure of goodness-of-fit is the PEV as defined in Harvey (1989). The PEV is the variance of the residuals in the steady state and so it corresponds to the variance of the disturbance of the reduced form ARIMA model. When the Kalman filter does not converge to the steady state, the finite PEV is used. For a univariate model, the PEV is denoted by $\tilde{\sigma}^2$, irrespective of whether or not it is the finite PEV.

10.4.2 Prediction error mean deviation

Another measure of fit is the mean deviation of the residuals, $v_t, t = d+1, \ldots, T$, defined by

$$md = \frac{\tilde{\sigma}}{T-d} \sum_{t=d+1}^{T} |v_t|$$

where the PEV $\tilde{\sigma}^2$ is defined in Harvey (1989). In a correctly specified model the reported ratio

$$\frac{2\tilde{\sigma}^2}{\pi \, md^2}$$

should be approximately equal to unity.

When it is indicated that the the series are modelled in logs, the relative measure of error is also given. This measure is easily computed as $100md$.

10.4.3 Coefficients of determination

In econometric modelling the traditional measure of goodness-of-fit is the coefficient of determination R^2. For a structural time series model this measure is defined by

$$R^2 = 1 - \frac{(T-d)\tilde{\sigma}^2}{\sum_{t=1}^{T}(y_t - \bar{y})^2},$$

where y_t is the original series and $\bar{y}$ is its sample mean. This coefficient of determination is most useful when the series appears to be stationary with no trend or seasonal.

When the series y_t shows trend movements, it is better to compare the PEV with the variance of first differences, $\Delta y_t = y_t - y_{t-1}$. This leads to

$$R_D^2 = 1 - \frac{(T-d)\tilde{\sigma}^2}{\sum_{t=2}^{T}(\Delta y_t - \overline{\Delta y})^2}.$$

where $\overline{\Delta y}$ is the sample mean of Δy_t. It should be noted that R_D^2 may be negative, indicating a worse fit than a simple random walk plus drift model.

For seasonal data with a trend, it is more appropriate to measure the fit against a random walk plus drift and fixed seasonals. This requires the sum of squares, SSDSM, obtained by subtracting the seasonal mean from Δy_t. The coefficient of determination is then

$$R_S^2 = 1 - \frac{(T-d)\tilde{\sigma}^2}{\text{SSDSM}},$$

which can again be negative.

10.4.4 Information criteria : AIC and BIC

The fit of different models can be compared on the basis of the PEV. However when models have different number of parameters, it is more appropriate to compare them using the Akaike information criterion (AIC) or the Bayes information criterion (BIC). These criteria are based on the formula

$$\log PEV + cm/T$$

where c is a constant and m is the number of parameters in θ plus the number of non-stationary components in the state vector. The AIC takes c as 2 and the BIC takes c as $\log T$.

10.5 Components

The 'Components' option facilitates the computation of filtered or smoothed estimates of the specified components in the model. The filtered estimates are based on the past observation and can be calculated as described in Koopman, Shephard and Doornik (1999). The smoothed estimates are based on all the observations. In what follows, the estimates are denoted by $\widetilde{x}$, where x is some component which can be estimated either way.

10.5.1 Series with components

This option plots the original series, y_t, with one of the following (as requested):

- the estimated trend, $\widetilde{\mu}_t$ (which includes interventions on the level and slope);
- the estimated trend plus cycle(s) plus AR(1), $\widetilde{\mu}_t + \widetilde{\psi}_t + \widetilde{v}_t$;
- the estimated trend plus cycle(s) plus AR(1) plus explanatory variables, $\widetilde{\mu}_t + \widetilde{\psi}_t + \widetilde{v}_t + \mathbf{x}_t' \widetilde{\boldsymbol{\delta}}_\mathbf{x}$, where the last part includes all the explanatory variables in the equation.

10.5.2 Detrended

The detrended series is $y_t - \widetilde{\mu}_t$ for $t = 1, \ldots, T$. The smoothed estimates are usually most appropriate for detrending. Note that level and slope interventions are included in the trend.

10.5.3 Seasonally adjusted

The seasonally adjusted series is $y_t - \widetilde{\gamma}_t$ for $t = 1, \ldots, T$. The smoothed estimate of the seasonal component is normally used for seasonal adjustment.

10.5.4 Individual seasonals

The estimated seasonal component $\widetilde{\gamma}_t$ is re-arranged into s different series, each corresponding to one of the seasons. The graph can be interpreted as a set of annual plots for the individual seasonal effects.

10.5.5 Data in logs

If the observations are in logarithms, the output of the various components may be amended in two ways: (1) 'Anti-logs' can be calculated; or (2) 'Modified anti-logs' can be calculated. These amendments do not apply in a similar way for all components. The following list gives the formal amendments:

- *Trend* - (1) $\exp(\widetilde{\mu}_t)$ and (2) $\exp(\widetilde{\mu}_t)$;
- *Growth rate (Slope)* - (1) $100s\,\widetilde{\beta}_t$ and (2) $100s\,\widetilde{\beta}_t$;
- *Seasonal, Cycle, AR(1) and Irregular* - (1) $\exp(\widetilde{a})$ and (2) $\exp(\widetilde{\mu}_t + \widetilde{a}) - \exp(\widetilde{\mu}_t)$, where $\widetilde{a}$ is a specific component;
- *Detrended* - (1) $\exp(y_t - \widetilde{\mu}_t)$ and (2) $\exp(y_t) - \exp(\widetilde{\mu}_t)$;
- *Seasonally adjusted* - (1) $\exp(y_t - \widetilde{\gamma}_t)$ and (2) $\exp(y_t - \widetilde{\gamma}_t)$.

10.6 Residuals

The residuals are the standardised one-step-ahead prediction errors or innovations and they are defined in Koopman, Shephard and Doornik (1999). The residual series are simply denoted by v_t for $t = d+1, \ldots, T$. For a correctly specified model with known variance parameters, the residuals are assumed to be $\text{NID}(0,1)$. Diagnostic statistics and graphs are the tools to validate this proposition.

10.6.1 Correlogram

The residual autocorrelation at lag τ is defined by

$$r_\tau = \frac{\sum_{t=\tau+d+1}^{T}(v_t - \bar{v})(v_{t-\tau} - \bar{v})}{\sum_{t=d+1}^{T}(v_t - \bar{v})^2}, \quad \text{where } \bar{v} = \frac{1}{T-d}\sum_{t=d+1}^{T} v_t. \tag{10.1}$$

The approximate standard error for r_τ is $1/\sqrt{T-d}$. The Durbin and Watson (1950) statistic, is given by

$$DW = \frac{\sum_{t=2+d}^{T}(v_t - v_{t-1})^2}{\sum_{t=1+d}^{T} v_t^2} \simeq 2\{1 - r(1)\}.$$

In a correctly specified model, DW is approximately $\text{N}(2, 4/T)$.

A general test for serial dependence is the portmanteau Box–Ljung Q-statistic, which is based on the first s residual autocorrelations; see Ljung and Box (1978). It is given by

$$Q(P, d) = T(T+2) \sum_{j=1}^{P} \frac{r_j^2}{T-j} \ .$$

and distributed approximately as χ_d^2 under the null, where d is equal to $P - n + 1$ and n is the number of parameters. Notice that the loss in the degrees of freedom d of $Q(P, d)$ takes account of the number of relative variance parameters. It does not take account of any lagged dependent variables and their presence necessitates further adjustment.

10.6.2 Periodogram and spectrum

These diagnostic graphs are defined in Harvey (1989). They are applied to v_t in exactly the same way. Note that the theoretical spectrum is a horizontal straight line for a white-noise series.

10.6.3 Cumulative statistics and graphs

The 'Cusum' and 'Cusum of squares' can be calculated using the residuals v_t for $t = d+1, \ldots, T$.

The 'Cumulative periodogram' is another cumulative graph which is useful since it smooths out the periodogram. The cumulative periodogram cp_j is defined as

$$cp_j = \frac{\sum_{i=1}^{j} p_i^2}{\sum_{i=1}^{n} p_i^2}$$

where p_i is the periodogram ordinate at i for $i = 1, \ldots, n$ and $n = [(T-d)/2]$. The cumulative periodogram cp_j is plotted against $j = 0, \ldots, n$. A useful diagnostic statistic is to measure the maximum deviation of the cp_j from the diagonal line connecting $cp_0 = 0$ and $cp_n = 1$; that is, $\max |cp_j - j/n|$ over $j = 1, \ldots, n-1$; see Durbin (1969). This measure is outputted on request in the Results window. Some typical critical values for this two-sided test are .44 for $T \simeq 80$, .40 for $T \simeq 100$ and .30 for $T \simeq 200$.

10.6.4 Distribution statistics

The Bowman and Shenton (1975) statistic for normality is calculated using the residuals v_t for $t = d+1, \ldots, T$.

10.6.5 Heteroskedasticity

A basic non-parametric test of heteroskedasticity is the two-sided $F_{h,h}$.-test,

$$H(h) = \sum_{t=T-h+1}^{T} v_t^2 / \sum_{t=d+1}^{d+1+h} v_t^2$$

where h is the closest integer to $(T - d)/3$. An increase of the variance in the last third of the sample will cause $H(h)$ to be large, while the reverse will cause it to be small. The heteroskedasticity test appears as part of the Summary statistics.

10.7 Auxiliary residuals

Auxiliary residuals are designed to detect unusual movements in a fitted time series model. These residuals are the smoothed estimates of the disturbances associated with the components irregular, level and slope. Koopman, Shephard and Doornik (1999) sets out how to calculate the standardised auxiliary residuals. It is very useful to plot them and look for any irregularities. If requested, all residuals which have absolute value exceeding two are written to the Results window.

As well as graphical representations of the auxiliary residuals, STAMP reports the standard distribution statistics for these residuals. Of course, these statistics cannot be used as standard moment tests because of the serial correlation in the auxiliary residuals. However these statistics can be corrected for serial correlation as in Harvey and Koopman (1992). In STAMP, the correction factors are calculated from the first $\max(\sqrt{T}, 20)$ sample autocorrelations of the auxiliary residuals.

10.8 Predictive testing

When the final date of the *state sample* is smaller than the final date of the *data sample*; that is, there are L observations, y_t, $t = T + 1, ..., T + L$ 'outside the sample', post-sample predictive testing may be carried out using all or some of these observations. However, when these final dates are the same, predictive testing within the sample is the only option; that is, predicting y_t for $t = T - l + 1, ..., T$. Of course, in the first situation, predictive testing within the sample is also possible.

When explanatory variables and/or interventions are specified in the model, the residuals used in predictive testing within the sample are the generalised recursive residuals defined in Harvey (1989). These residuals are denoted by w_t for $t = d^* + 1, ..., T$ where $d^* = d + k$ and k is the number of explanatory variables in the model (excluding any variables which do not appear after time $T - L$). Note that w_t is zero when an intervention enters into the model. The error bands for the residuals are based on twice the RMSE. The CUSUM is calculated for these residuals if requested.

The Chow test within the sample is constructed as

$$Chow = \frac{T-l-d^*}{l} \sum_{t=T-l+1}^{T} w_t^2 \bigg/ \sum_{t=d^*+1}^{T-l} w_t^2.$$

which is approximately distributed as $F(l, T-l-d^*)$.

When post-sample predictive testing is carried out, the estimates of the coefficients of the explanatory variables and/or interventions remain the same as they were at the end of the state sample. The corresponding standardised residuals are denoted by v_t for $t = T+1, \ldots, L$. The post-sample predictive 'failure' test statistic is then

$$pft = \sum_{j=1}^{L} v_{T+j}^2,$$

which is approximately distributed as χ_L^2.

The CUSUM t-test is also outputted in the form

$$cusumt = L^{-1/2} \sum_{j=1}^{L} v_{T+j},$$

which is approximately distributed as a t distribution with $T - L - d^*$ degrees of freedom.

The extrapolative residuals, which are defined in Harvey (1989), can be used to assess the predictive performance of the model and make comparisons with rival models. If $\hat{v}_{T+j|T}$, $j = 1, \ldots, L$, denotes the j-step ahead extrapolative residual in the post-sample period, the *extrapolative sum of squares* is given by

$$ess(T, L) = \sum_{j=1}^{L} \hat{v}_{T+j|T}^2.$$

This statistic can also be generated using residuals within the sample. A similar statistic based on the sum of absolute values is defined as

$$esa(T, L) = \sum_{j=1}^{L} |\hat{v}_{T+j|T}|.$$

10.9 Forecast

The forecasts of the series y_t and the components are generated as described in Koopman, Shephard and Doornik (1999). The forecast error bounds are based on one RMSE.

For example, the forecasts of a specific series $\widehat{y}_{T+j}$ are obtained from a particular element of $\widehat{\mathbf{y}}_{T+j|T}$, for $j = 1, \ldots, L$, and the RMSE $\widehat{r}_{T+j}$ is the square root of the corresponding diagonal element of the matrix $\widehat{\sigma}^2 \widehat{\mathbf{F}}_{T+j|T}$. When the data are in logs, the anti-log transformation (exp) can be calculated for the forecasts. Their corresponding RMSEs are then based on

$$\widehat{r}^*_{T+j} = \exp(\widehat{y}_{T+j} + \widehat{r}_{T+j}) - \exp(\widehat{y}_{T+j}), \quad j = 1, \ldots, L.$$

The anti-log forecasts adjusted for bias are calculated by

$$\widehat{y}^*_{T+j} = \exp(\widehat{y}_{T+j} + 0.5 \widehat{r}^2_{T+j}), \quad j = 1, \ldots, L.$$

Appendix A1
STAMP Batch Language

STAMP is mostly menu-driven for ease of use. To add flexibility, certain functions can be accessed through entering commands. The syntax of these commands is described in this chapter. They can be added to the general list of Batch commands provided by OxMetrics.

STAMP allows models to be formulated, estimated and evaluated through batch commands. Such commands are entered in OxMetrics. Certain commands are intercepted by OxMetrics, such as those for loading and saving data, as well as blocks of algebra code. The remaining commands are then passed on to the active module, which is STAMP in this case. This section gives an alphabetical list of the STAMP batch language statements. There are two types of batch commands: function calls (with or without arguments) terminated by a semicolon, and commands, which are followed by statements between curly brackets.

Anything between /* and */ is considered comment. Note that this comment cannot be nested. Everything following // up to the end of the line is also comment.

OxMetrics allows you to save the current model as a batch file, and to rerun saved batch files. If a model has been created interactively, it can be saved as a batch file for further editing or easy recall in a later session. This is also the most convenient way to create a batch file.

If an error occurs during processing, the batch run will be aborted and control returned to OxMetrics. A warning or out of memory message will have to be accepted by the user (press Enter), upon which the batch run will resume.

Batch is activated in OxMetrics via **Model**/Batch or just by pressing Alt+b. The batch dialog appears with default code displayed. The code reflects the current model in STAMP. For example,

Appendix A1 STAMP Batch Language

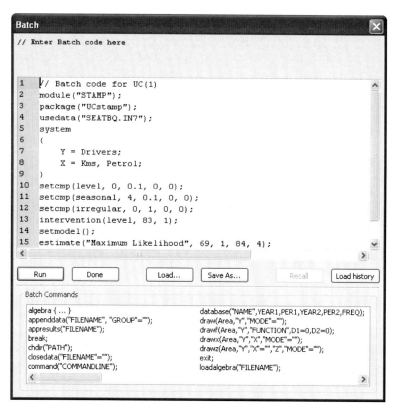

In the following list, function arguments are indicated by *words*, whereas the areas where statement blocks are expected are indicated by .... Examples follow the list of descriptions. For terms in double quotes, the desired term must be substituted and provided together with the quotes.

`estimate()();`

Starts the estimation process as described in §8.2.2 and §9.6.

`forecast(`*nrfor, incr, option*`);`

Prints *nrfor* ahead forecasts of dependent variables. When explanatory variables are included in the model, the *incr* value is the increment value for all exogenous variables. Otherwise, no value for *incr* should be given. When a correction for bias is required due to the log-transformation, the *option* must be set to logbias. Otherwise, no value is required for *option*.

`intervention(`*component, year, period*`);`

The *component* argument specifies the type of intervention to be included in the model:

- include outlier: `outlier`;
- include structural break: `level`;

- include slope change: `slope`;
- no intervention in the model: `off`.

The *year* and *period* indicates the start of the intervention. When *year* and *period* refer to a non-existing date within the sample, the intervention will not enter the model.

system { Y=...; X=...; }

Formulate the model, consisting of the following components:

Y dependent variables;
X explanatory (including lagged dependent) variables;

The variables listed are separated by commas, the base names (that is, name excluding lag length) for Y and X must be in the database. If the variable names are not a valid token, the name must be enclosed in double quotes.

store(*type*, *variable*);

Use this command to store prediction residuals, smoothed components and smoothed auxiliary residuals into the database, the default name is used. The *type* must be one of: `residuals, components, auxresiduals`. The *variable* must be one of: `irreg, level, slope`.

When *type* is `residuals`, no *variable* need to be specified.

When *type* is `components`, the *variable* can also be one of: `trend, seaso, cycle, detrend, seasadj, trendx`. The variable `cycle` refers to the sum of smoothed components associated with autoregressive and the three cycles. The variable `trend` is the same as the smoothed level and `trendx` refers to the sum of smoothed level and all estimated explanatory variables.

setcmp(cmp, order, variance, par1, par2);

Introduces an unobserved component in the model. The possibilities are:

- `setcmp(level, -, 1.0, -, -)`; where - refers to "not a relevant number" and `1.0` is the variance.
- `setcmp(slope, 2, 1.0, -, -)`; where 2 is the order of the trend.
- `setcmp(seasonal, 4, 1.0, -, -)`; where 4 is the length of the seasonal.
- `setcmp(irregular, -, 1.0, -, -)`;
- `setcmp(cycle, 2, 1.0, 5, 0.9)`; where 2 is the order of the cycle, 5 is the period and `0.9` is the discounting factor of the cycle.
- `setcmp(ar, 1, 1.0, 0.8, -)`; where 1 is the order of the autoregressive process, AR(1) and `0.8` is the autoregressive coefficient.
- `setcmp(ar, 2, 1.0, 0.8, 0.2)`; where 2 is the order of the autoregressive process, AR(2), `0.8` is the first autoregressive coefficient and `0.2` is the second autoregressive coefficient.

setmodel();

This call is required to activate all specified unobserved components in the model. This call must always come after the last setcmp statement.

teststate();

Prints the estimated final state and the estimated regression and intervention vector together with the corresponding test statistics. The output depends on the specified model.

testsummary();

Prints the estimation and diagnostic summary reports.

We finish with an annotated example using most commands. To run this file, we assume that OxMetrics is loaded with SEATBQ.IN7, and that STAMP has been started.

```
// Batch code for UC( 1)
module("STAMP");
package("UCstamp");
usedata("SEATBQ.IN7");
system
{
    Y = Drivers;
    X = Kms, Petrol;
}
setcmp(level, 0, 0.1, 0, 0);
setcmp(seasonal, 4, 0.1, 0, 0);
setcmp(irregular, 0, 1, 0, 0);
intervention(level, 83, 1);
setmodel();
estimate("Maximum Likelihood", 69, 1, 84, 4);
```

This Batch code produces the output:

```
UC( 1) Modelling Drivers by Maximum Likelihood (using SEATBQ.IN7)
    The selection sample is: 69(1) - 84(4)
    The model is:  Y = Level + Seasonal + Irregular + Expl vars + Interv

Log-Likelihood is 136.987 (-2 LogL = -273.975).
Prediction error variance is 0.00686482

Summary statistics
    std.error       0.082854
    Normality       5.2573
    H(19)           0.56844
    r(1)            0.16536
    r(6)            0.046724
    DW              1.6354
    Q(6,4)          10.348
    Rs^2            0.49962

Variances of disturbances.

Component               Value    (q-ratio)
Level               0.00035069  ( 0.0715)
```

```
Seasonal                 1.2587e-005 ( 0.0026)
Irregular                0.0049063 ( 1.0000)

State vector analysis at period 84(4)
 - level is 4.65234 with stand.err 1.7011.
 - joint seasonal chi2 test is 72.2287 with 3 df.
 - seasonal effects are
period      value    stand.err
     1   -0.073881   0.026365
     2   -0.142151   0.028420
     3   -0.013761   0.027807
     4    0.229794   0.031621

Regression effects in final state at time 84(4)

                 Coefficient   stand.err    t-value      prob
Kms                -0.21887     0.05351    -4.09054  [0.00014]
Petrol              0.22469     0.17688     1.27030  [0.20914]
Level break 83. 1  -0.26824     0.12211    -2.19677  [0.03212]
```

References

Anderson, B. D. O. and Moore, J. B. (1979). *Optimal Filtering*. Englewood Cliffs: Prentice-Hall.

Ansley, C. F. and Kohn, R. (1986). A note on reparameterizing a vector autoregressive moving average model to enforce stationarity, *J. Statistical Computation and Simulation*, **24**, 99–106.

Balke, N. S. (1993). Detecting level shifts in time series, *Journal of Business and Economic Statistics*, **11**, 81–92.

Baxter, M. and King, R. (1999). Measuring business cycles: approximate band-pass filters for economic time series, *Rev. Economics and Statistics*, **81**, 575–93.

Bowman, K. O. and Shenton, L. R. (1975). Omnibus test contours for departures from normality based on $\sqrt{b_1}$ and b_2, *Biometrika*, **62**, 243–50.

Box, G. E. P. and Jenkins, G. M. (1970). *Time Series Analysis: Forecasting and Control*. San Francisco, CA: Holden-Day.

Bruce, A. G. and Martin, R. D. (1989). Leave–k–out diagnostics for time series (with discussion), *Journal of the Royal Statistical Society, Series B*, **51**, 363–424.

Bryan, A. C. and Cecchetti (1994). Measuring core inflation, In Mankiw, N. G. (ed.), *Monetary Policy*, pp. 195–215. Chicago: University of Chicago Press.

Bryson, A. E. and Ho, Y. C. (1969). *Applied Optimal Control*. Massachusetts: Blaisdell.

Busetti, F. (2006). Tests of seasonal integration and cointegration in multivariate unobserved components models, *J. Applied Econometrics*, **21**, 419–38.

Carvalho, V., Harvey, A. C. and Trimbur, T. (2007). A note on common cycles, common trends and convergence, *J. Business and Economic Statist.*, **23**, 12–20.

Chan, W. Y. T. and Wallis, K. F. (1978). Multiple time series modelling: another look at the mink-muskrat interactions, *Applied Statistics*, **27**, 168–75.

Christiano, L. and Fitzgerald, T. (2003). The band-pass filter, *International Economic Review*, **44**, 435–65.

Chu-Chun-Lin, S. and de Jong, P. (1993). A note on fast smoothing, Unpublished: University of British Columbia.

Clark, P. K. (1989). Trend reversion in real output and unemployment, *Journal of Econometrics*, **40**, 15–32.

Cogley, T. and Sargent, T. (2007). Inflation-Gap persistence in the US, Mimeo.

Commandeur, J. J. F. and Koopman, S. J. (2007). *An Introduction to State Space Time Series Analysis*. Oxford: Oxford University Press.

Cramer, J. S. (1986). *Econometric Applications of Maximum Likelihood Methods*. Cambridge: Cambridge University Press.

de Jong, P. (1988). The likelihood for a state space model, *Biometrika*, **75**, 165–169.

de Jong, P. (1989). Smoothing and interpolation with the state space model, *Journal of the American Statistical Association*, **84**, 1085–8.

de Jong, P. (1991). The diffuse Kalman filter, *Annals of Statistics*, **19**, 1073–1083.

de Jong, P. and Chu-Chun-Lin, S. (1994). Fast likelihood evaluation and prediction for nonstationary state space models, *Biometrika*, **81**, 133–142.

den Butter, F. A. G. and Mourik, T. J. (1990). Seasonal adjustment using structural time series models: an application and comparison with the X–11 method, *Journal of Business and Economic Statistics*, **8**, 385–94.

Doornik, J. A. and Hansen, H. (1994). An omnibus test of univariate and multivariate normality, Unpublished paper: Nuffield College, Oxford.

Durbin, J. (1969). Tests for serial correlation in regression analysis based on the periodogram of least-squares residuals, *Biometrika*, **56**, 1–15.

Durbin, J. and Koopman, S. J. (2001). *Time Series Analysis by State Space Methods*. Oxford: Oxford University Press.

Durbin, J. and Watson, G. S. (1950). Testing for serial correlation in least squares regression, I, *Biometrika*, **37**, 409–428.

Engle, R. F. and Kozicki, S. (1993). Testing for common features (with discussion), *Journal of Business and Economic Statistics*, **11**, 369–395.

Fuller, W. A. (1996). *Introduction to Time Series* 2nd edition. New York: John Wiley.

Ghysels, E., Harvey, A. C. and Renault, E. (1996). Stochastic volatility, In Rao, C. R. and Maddala, G. S. (eds.), *Statistical Methods in Finance*, pp. 119–191. Amsterdam: North-Holland.

Gill, P. E., Murray, W. and Wright, M. H. (1981). *Practical Optimization*. New York: Academic Press.

Gomez, V. (2001). The use of Butterworth filters for trend and cycle estimation in economic time series, *J. Business and Economic Statist.*, **19**, 365–73.

Hansen, P. R. and Lunde, A. (2006). Realized variance and market microstructure noise (with discussion), *J. Business and Economic Statist.*, **24**, 127–218.

Harvey, A. C. (1989). *Forecasting, Structural Time Series Models and the Kalman Filter*. Cambridge: Cambridge University Press.

Harvey, A. C. (1993). *Time Series Models* 2nd edition. Hemel Hempstead: Harvester Wheatsheaf.

Harvey, A. C. (2001). Testing in unobserved components models, *J. of Forecasting*, **20**, 1–19.

Harvey, A. C. (2006). Forecasting with unobserved components time series models, In Elliot, G., Granger, C. W. and Timmermann, A. (eds.), *Handbook of Economic Forecasting*. Amsterdam: North Holland.

Harvey, A. C. and Durbin, J. (1986). The effects of seat belt legislation on British road casualties: A case study in structural time series modelling, *Journal of the Royal Statistical Society, Series B*, **149**, 187–227.

Harvey, A. C., Henry, B., Peters, S. and Wren-Lewis, S. (1986). Stochastic trends in dynamic regression models: an application to the employment-output equation, *Economic Journal*, **96**, 975–985.

Harvey, A. C. and Jaeger, A. (1993). Detrending, stylised facts and the business cycle, *Journal of Applied Econometrics*, **8**, 231–47.

Harvey, A. C. and Koopman, S. J. (1992). Diagnostic checking of unobserved components time series models, *Journal of Business and Economic Statistics*, **10**, 377–389.

Harvey, A. C. and Koopman, S. J. (2000). Signal extraction and the formulation of unobserved components models, *Econometrics Journal*, **3**, 84–107.

Harvey, A. C. and Proietti, T. (2005). *Readings in Unobserved Components Models*. Oxford: Oxford University Press.

Harvey, A. C., Ruiz, E. and Shephard, N. (1994). Multivariate stochastic variance models, *Review of Economic Studies*, **61**, 247–264.

Harvey, A. C. and Shephard, N. (1993). Structural time series models, In Maddala, G. S., Rao, C. R. and Vinod, H. D. (eds.), *Handbook of Statistics, Volume 11*. Amsterdam: Elsevier Science Publishers B V.

Harvey, A. C. and Trimbur, T. (2003). Generalised model-based filters for extracting trends and cycles in economic time series, *Rev. Economics and Statistics*, **85**, 244–55.

Harvey, A. C., Trimbur, T. and van Dijk, H. K. (2007). Trends and cycles in economic time series: a Bayesian approach, *J. Econometrics*, **140**, 618–49.

Hodrick, R. J. and Prescott, E. C. (1980). Postwar U.S. business cycles: an empirical investigation, Discussion paper 451, Carnegie-Mellon University.

Hull, J. and White, A. (1987). The pricing of options on assets with stochastic volatilities, *Journal of Finance*, **42**, 281–300.

Hylleberg, S., Engle, R. F., Granger, C. W. J. and Yoo, B. S. (1990). Seasonal integration and cointegration, *J. Econometrics*, **44**, 215–38.

Kane, R. P. and Trivedi, N. B. (1986). Are droughts predictable, *Climate Change*, **8**, 208–23.

Kim, S., Shephard, N. and Chib, S. (1998). Stochastic volatility: likelihood inference and comparison with ARCH models, *Review of Economic Studies*, **65**, 361–393.

King, R., Plosser, C., Stock, J. H. and Watson, M. W. (1991). Stochastic trends and economic fluctuations, *American Economic Review*, **81**, 819–840.

Kitagawa, G. and Gersch, W. (1984). A smoothness prior — state space modeling of time series with trend and seasonality, *Journal of the American Statistical Association*, **79**, 378–89.

Kohn, R. and Ansley, C. F. (1989). A fast algorithm for signal extraction, influence and cross-validation, *Biometrika*, **76**, 65–79.

Koopman, S. J. (1993). Disturbance smoother for state space models, *Biometrika*, **80**, 117–126.

Koopman, S. J., Harvey, A. C., Doornik, J. A. and Shephard, N. (2007). *STAMP 8.0: Structural Time Series Analyser, Modeller and Predictor*. London: Timberlake Consultants.

Koopman, S. J. and Shephard, N. (1992). Exact score for time series models in state space form, *Biometrika*, **79**, 823–6.

Koopman, S. J., Shephard, N. and Doornik, J. A. (1999). Statistical algorithms for models in state space form using SsfPack 2.2, *Econometrics Journal*, **2**, 113–66. http://www.ssfpack.com/.

Ljung, G. M. and Box, G. E. P. (1978). On a measure of lack of fit in time series models, *Biometrika*, **66**, 67–72.

Mahieu, R. and Schotman, P. (1998). An empirical application of stochastic volatility models, *Journal of Applied Econometrics*, **16**, 333–59.

Maravall, A. (1985). On structural time series models and the characterization of components, *Journal of Business and Economic Statistics*, **3**, 350–355.

Mills, T. C. (1993). *Time Series Methods for Finance*. Cambridge: Cambridge University Press.

Morrettin, P. A., Mesquita, A. R. and Rocha, J. G. C. (1985). Rainfall in Fortaleza in Brazil revisted, In Anderson, O. D., Robinson, E. A. and Ord, J. K. (eds.), *Time Series Analysis: Theory and Practice 6*. Amsterdam: North Holland.

Newton, J. H., North, G. R. and Crowley, T. J. (1991). Forecasting global ice volume, *Journal of Time Series Analysis*, **12**, 255–265.

Nyblom, J. and Harvey, A. C. (2001). Testing against smooth stochastic trends, *J. Applied Econometrics*, **16**, 415–29.

Nyblom, J. and Makelainen, T. (1983). Comparison of tests of for the presence of random walk coefficients in a simple linear models, *Journal of the American Statistical Association*, **78**, 856–64.

Prest, A. R. (1949). Some experiments with demand analysis, *Review of Economics and Statistics*, **31**, 33–49.

Sandmann, G. and Koopman, S. J. (1998). Estimation of stochastic volatility models via Monte Carlo maximum likelihood, *Journal of Econometrics*, **87**, 271–301.

Shephard, N. (1993). Maximum likelihood estimation of regression models with stochastic trend components, *Journal of the American Statistical Association*, **88**, 590–595.

Shephard, N. (1996). Statistical aspects of ARCH and stochastic volatility, In Cox, D. R., Hinkley, D. V. and Barndorff-Nielsen, O. E. (eds.), *Time Series Models in Econometrics, Finance and Other Fields*, pp. 1–67. London: Chapman & Hall.

Shephard, N. (2005). *Stochastic Volatility: Selected Readings*. Oxford: Oxford University Press.

Shumway, R. H. and Stoffer, D. (1982). An approach to time series smoothing and forecasting using the EM algorithm, *Journal of Time Series Analysis*, **3**, 253–264.

Thisted, R. A. (1988). *Elements of Statistical Computing*. New York: Chapman & Hall.

Tunnicliffe-Wilson, G. (1989). On the use of marginal likelihood in time series model estimation, *Journal of the Royal Statistical Society, Series B*, **51**, 15–27.

Watson, M. W. and Engle, R. F. (1983). Alternative algorithms for the estimation of dynamic factor, mimic and varying coefficient regression, *Journal of Econometrics*, **23**, 385–400.

West, M. and Harrison, J. (1989). *Bayesian Forecasting and Dynamic Models*. New York: Springer-Verlag.

West, M. and Harrison, J. (1997). *Bayesian Forecasting and Dynamic Models* 2 edition. New York: Springer-Verlag.

Author Index

Anderson, B. D. O. 169, 180
Ansley, C. F. 185, 188, 191

Balke, N. S. 16, 69
Baxter, M. 64, 110
Bowman, K. O. 209
Box, G. E. P. 18, 19, 53, 209
Bruce, A. G. 19
Bryan, A. C. 116
Bryson, A. E. 185
Busetti, F. 101

Carvalho, V. 101, 124
Cecchetti 116
Chan, W. Y. T. 16, 91
Chib, S. 125
Christiano, L. 110
Chu-Chun-Lin, S. 176, 181, 186
Clark, P. K. 106
Cogley, T. 116
Commandeur, J. J. F. 4, 31
Cramer, J. S. 197
Crowley, T. J. 19, 67

de Jong, P. 169, 175, 176, 181, 182, 185, 186, 188
den Butter, F. A. G. 128
Doornik, J. A. xix, 4, 13, 201, 203–205, 207, 208, 210, 211
Durbin, J. 4, 17, 19, 75, 103, 136, 208, 209

Engle, R. F. 101, 186

Fitzgerald, T. 110
Fuller, W. A. 78, 87

Gersch, W. 128
Ghysels, E. 125
Gill, P. E. 197
Gomez, V. 55
Granger, C. W. J. 101

Hansen, H. 201
Hansen, P. R. 133
Harrison, J. 4, 19, 175

Harvey, A. C. 4, 13, 14, 17, 19, 52, 56, 63, 64, 66, 70, 72, 75, 82, 87, 95, 101–103, 106, 109, 110, 114, 120, 124, 125, 128, 129, 135, 136, 160, 169, 175, 188, 205, 206, 209–211
Henry, B. 19, 87
Ho, Y. C. 185
Hodrick, R. J. 128
Hull, J. 125
Hylleberg, S. 101

Jaeger, A. 128, 129
Jenkins, G. M. 18, 19, 53

Kane, R. P. 16, 65
Kim, S. 125
King, R. 17, 64, 110
Kitagawa, G. 128
Kohn, R. 185, 188, 191
Koopman, S. J. xix, 4, 13, 19, 31, 70, 72, 82, 109, 125, 160, 169, 185–188, 203–205, 207, 208, 210, 211
Kozicki, S. 101

Ljung, G. M. 209
Lunde, A. 133

Mahieu, R. 125
Makelainen, T. 56
Maravall, A. 128
Martin, R. D. 19
Mesquita, A. R. 65
Mills, T. C. 16
Moore, J. B. 169, 180
Morrettin, P. A. 65
Mourik, T. J. 128
Murray, W. 197

Newton, J. H. 19, 67
North, G. R. 19, 67
Nyblom, J. 56, 95

Peters, S. 19, 87
Plosser, C. 17
Prescott, E. C. 128

Prest, A. R. 17
Proietti, T. 4

Renault, E. 125
Rocha, J. G. C. 65
Ruiz, E. 102, 125

Sandmann, G. 125
Sargent, T. 116
Schotman, P. 125
Shenton, L. R. 209
Shephard, N. xix, 4, 13, 102, 125, 176, 187, 203–205, 207, 208, 210, 211
Shumway, R. H. 186
Stock, J. H. 17
Stoffer, D. 186

Thisted, R. A. 197
Trimbur, T. 63, 64, 101, 106, 110, 120, 124
Trivedi, N. B. 16, 65
Tunnicliffe-Wilson, G. 176

van Dijk, H. K. 120, 124

Wallis, K. F. 16, 91
Watson, G. S. 208
Watson, M. W. 17, 186
West, M. 4, 19, 175
White, A. 125
Wren-Lewis, S. 19, 87
Wright, M. H. 197

Yoo, B. S. 101

Subject Index

Amplitude 62, 91, 203, 205,
    Also see Cycle
ARCH *see* Autoregressive conditional heteroskedasticity, Stochastic volatility
ARIMA *see* Autoregressive integrated moving average
Augmented disturbance smoothing 186
Augmented Kalman filter 180, 182, 184, 196
Autocorrelation function 61, 71, 90,
    Also see Correlogram
Autoregression 66, 92
Autoregressive 175
Autoregressive conditional heteroskedasticity (ARCH) 125
Autoregressive integrated moving average (ARIMA) 53, 69, 87, 205
Auxiliary residuals 8, 44, 70, 161, 185, 188, 210, 215
— graphics 71, 87, 161

Bandpass filter 63
Basic structural model 34, 57, 76, 88, 128, 141
    Multivariate 89
Batch 9, 213
— file (.FL) 213
    Language 213
    Variable names 215
Bayesian 4
Bowman-Shenton statistic 154, 209
Box–Jenkins method 49,
    Also see ARIMA
Box–Ljung statistic 36, 80, 153, 201, 209
Broyden-Fletcher-Goldfarb-Shanno (BFGS) estimation 148, 191, 192, 198, 199
Bureau of the Census X-11 66

Cholesky decomposition 94, 95, 102, 172, 190, 194
Citation 13
Coefficient of Determination 201, 206
Cointegrating vector 97
Cointegration 93, 95, 114
    Seasonal 101
Common cycles 101, 174
Common factors 88, 93, 172, 174, 178
    Balanced levels 97
    Common cycles 101
    Common seasonals 101
    Common trends 99
Common seasonals 101, 174
Common trends 94, 99, 172–174
Components 215
— graphics 39, 51, 70, 81, 91, 131, 152, 158
    Autoregressive 5, 6, 55, 63, 66, 81, 87, 92, 106, 141, 169, 170, 172, 189, 191, 200, 202, 204
    Cycle 6, 87, 90, 113, 141, 177, 204, 205, 215
        Similar cycles 120, 169
        Stochastic 66, 170, 202
        Trigonometric 106
    Irregular 6, 34, 78, 111
    Seasonal 6, 56, 89, 101, 113, 118, 119, 128, 159, 169, 179, 204, 208, 215
        Dummy variable 204
        Fixed 49, 170, 174, 179, 190
        Stochastic 34
        Trigonometric 5, 89, 101, 169, 203
    Selection 33, 49, 140
    Trend 6, 56, 159, 170, 171, 176, 178, 207, 215
        Fixed 49, 170, 174, 179, 190
        Level 78, 203, 205, 207, 210
        Local level 140
        Slope 78, 207, 210

SUBJECT INDEX

Stochastic   34, 140, 169, 173, 177, 179
Concentrated likelihood   182
Control group   103
Convergence   153, 192, 198, 199, 201
Correlation matrix   6, 88, 95, 113, 123, 181
Correlogram   42, 53, 61, 160, 208
Cumulative periodogram   209
CUSUM   209, 211
CUSUMQ   209
Cycle   60, 62, 63, 65, 66, 169–172, 174, 177, 178, 189, 190, 202–205, 208
— higher order   63
— plus noise   63
— plus noise model   60
Multiple cycle   64
Stochastic —   60, 62, 63, 66, 170

Data
— loading   21
— storage   10
— transformations   25
Spreadsheet files   10
Data set
Airline — (AIRLINE)   18
Brazilian rainfall — (RAINBRAZ)   16, 60, 64
Employment and output — (EMPL)   18, 87
Energy — (ENERGY)   14, 21, 31, 47, 87
Exchange rate — (EXCH)   15, 125
Ice volume — (ICEVOL)   19, 67
Interest rate — (INTEREST)   16, 185
Latin American exports — (LAXQ)   19, 87
Mink and Muskrat — (MINKMUSK)   16, 90
Nile — (NILE)   16, 69–71, 86, 87
Purse — (PURSE)   19
Seat belt monthly — (SEATBELT)   16, 75, 87
Seat belt quarterly — (SEATBQ)   16, 95, 103, 136
Spirits — (SPIRIT)   17, 78, 83, 87
Telephone calls — (TELEPHON)   17
Trade price — (TRADEPRICE.IN7)   133
UK consumption, income and prices — (UKCYP)   17, 57, 89

US income, consumption, investment, money and prices — (USmacro07)   18, 107, 113
US income, consumption, investment, money and prices — (USYCIMP)   17, 50, 64
Detrending   9, 91, 127, 128, 135, 159, 185, 187, 207, 215
Diagnostic checking   9, 38, 42, 49, 152, 153, 160, 179, 185, 201, 205
Dialog
Multiple selection list box   11, 24
Differencing   170
First difference   87
Diffuse
— distributions   186, 196
— initial conditions   175, 176
— likelihood   182, 187
Disturbance smoothing   185
Documentation   3, 12
Doornik-Hansen statistic   201
Drift   140, 170, 206
Dummy variable   68
Durbin-Watson statistic   78, 80, 201

EM algorithm   185–187
Estimated variances   36, 80, 89, 188
Excel   9, 10, 21
Explanatory variables   6, 75, 77, 82, 102, 138, 164, 171, 175, 204, 205, 210, 215
Extrapolation   163
Extrapolative residuals   211

Factor loadings   94, 97, 100, 172, 191, 195
Factor rotations   102
Failure chi2 test   76
Filtering   5, 111, 184, 189, 196, 207
Final state vector   38, 91, 100, 202, 203
Finite difference approximation   199
Forecasting   5, 6, 8, 10, 38, 47, 59, 70, 83, 152, 164, 188, 214
— explanatory variables
    incremental change   84
    manual input   85
    model-based   86

Gain functions   8, 110

SUBJECT INDEX

— graphics 107
Generalised least squares 180, 181, 184
Goodness of fit 36, 49, 51, 80, 153, 154, 201–203, 205, 206
Graph
  — printing 25
  — saving 25

Help 12
Hessian 192, 199, 203
Heteroskedasticity 80, 154, 201, 210
  Stochastic volatility 125
Higher order cycle 6, 63, 106
Hodrick-Prescott filter 128

IN7/BN7 files 10, 21
Innovations 179, 181, 184, 202, 205, 208
Interventions 6, 68, 71, 86, 87, 102, 103, 141, 171, 214
  — selection 69, 143

Kalman filter 7, 53, 125, 179–181, 196, 205
  Initialisation 180
  Steady-state 180, 182, 183, 201, 205

Lags 82, 118, 138
Likelihood kernel 193, 194, 200, 201
Likelihood ratio 99, 101
Line search 198
List box
  Multiple-selection — 11, 24
Local level model 50, 94, 97, 171, 172, 182
  Multivariate 89
Local linear trend 53, 55, 86, 106, 173, 178
Lotus 9, 10, 21

Maximum likelihood estimation 7, 34, 107, 125, 147, 181, 182, 186, 189, 197, 200
Measurement equation 86, 175, 176, 186
Missing values 5, 130
Model
  — estimation 3, 35, 51, 61, 70, 79, 90, 97, 98, 108, 113, 130, 133, 147, 153, 154, 157, 200, 201, 214, 216
  — evaluation 3, 38, 80, 114, 123, 126, 151
  — formulation 3, 31, 81, 88, 90, 94, 97, 103, 106, 113, 120, 126, 137, 142, 216

Variance matrix structures 89
  More written output 38, 57, 152, 155
Multi-step prediction 45, 50
Multivariate 184
  — diffuse likelihood 183
  — model 3, 5, 88–105, 119, 171, 173, 178, 183, 190, 194

Newton-Raphson 198
Non-parametric 210
Nonlinear 193, 197, 198
Normality test 188
Numerical optimisation 9, 36, 141, 148, 191, 197, 200, 201

Outliers 5, 7, 68, 70, 82, 87, 154, 161, 185, 188, 214
OxMetrics 3, 4, 20, 27, 152, 213
  — graphics 23
  Modules menu 20
  Calculator 25
  Graphics 31
  Model/Tail Probability 36
OxMetrics Data File (IN7/BN7) 10

P-value 36
PcGive 19
Periodogram 65, 209
Portmonteau statistic 209
Post-intervention predictive test 76, 87
Post-sample predictive test 48, 210, 211
Prediction 5, 44, 50, 76, 81, 162, 164, 179–181, 183, 202, 205, 206, 208, 215
  — graphics 44, 76, 163, 189
  Multi-step 45, 163
Prediction error decomposition 179, 181
Prediction error variance 154, 183, 201, 206
Predictive failure test 163

Q statistic *see* Box–Ljung test
Q-ratio 7, 53, 91, 107, 182, 195, 202
Quasi Newton 191, 193, 198
Quasi-maximum likelihood estimation 125

Random walk 5, 52, 97, 99, 106, 116, 119, 133, 140, 142, 170, 206
  —plus white noise 50, 71
Rank 94

SUBJECT INDEX

Reduced rank 6, 88, 94, 99, 190
Regression 6, 77, 80, 87, 175, 176, 179, 180, 183, 184, 196, 197, 203, 204
— fixed 142
— time-varying 142
Residuals 57, 60, 70, 81, 160, 161, 179, 180, 184, 185, 188, 189, 197, 205, 206, 208–211
— graphics 42, 152, 160
Auxiliary — graphics 71, 161
Innovations 179, 181, 184, 202, 205, 208
Results
— storage 12
Diagnostic summary report 36
Estimation report 36
Root mean square error 17, 45, 52, 179, 181, 188, 203, 210, 212

Sample autocorrelations 210
Sample period 11
Score vector 185–187
Seasonal adjustment 59, 127, 128, 159, 185, 187, 207, 215
Seasonal component 56, 57, 59, 207
Fixed 60, 63, 65
Seasonality 56, 58, 60, 70, 101, 174, 204
Fixed 56, 101
Seemingly unrelated time series equations (SUTSE) 88, 94, 171, 172, 174, 178
Serial correlation 36, 56, 57, 60, 70, 72, 80, 153, 185, 188, 201, 210
Signal extraction 9, 40, 53, 134, 152, 158, 185
Signal to noise ratio 52
Smoothing 5, 49, 53, 111, 125, 179, 185, 186, 188, 189, 196, 207
Component smoothing 187, 215
Disturbance smoothing 185
State smoothing 188
Spectral gain function 160
Spectrum 5, 90, 160, 209
Spline 5, 128, 142
Spurious cycles 129
STAMP menus
Intervention 69
Matrix editor 86
Model menu 31, 136
Algebra 31
Batch 213
Calculator 25, 31
Components 33
Estimate 35, 141
Formulate 11, 32, 141
Test 11
Progress 151
Test menu 38, 148, 200, 202
Components graphics 39, 152
Forecasting 47
More written output 38, 152
Prediction graphics 44
Residuals graphics 42
Starting STAMP 20
State space 4, 125, 169, 175, 177, 192,
Also see Structural time series model
Measurement equation 86, 175, 176, 186
State vector 176–181, 183–185, 187, 188, 203
Structural time series model 203, 206
Transition equation 175, 186
State vector 176–181, 183–185, 187, 188
Stationarity 53, 116
Cycle 60, 62, 63, 65, 66, 87, 90, 95, 169–172, 174, 177, 178, 189, 190, 202–205, 208
Step length 198, 199
Stochastic volatility 125, 135
Structural breaks 5, 7, 16, 68, 70, 82, 176, 185, 214
Structural time series model 3, 56, 66, 177, 178, 180, 183, 203, 206
Basic structural model 34, 57, 88, 141
Estimation 79, 147, 153, 178–180, 182, 183, 185, 189–193, 195, 200, 201
Explanatory variables 68, 77–211
Local level 50, 94, 97, 140, 170–172, 182
Local linear trend 53, 55, 173
Reduced form 87, 205
Seasonality 174, 204
SUTSE *see* Seemingly unrelated time series equations

Transition equation 175, 186
Trend 50, 56, 169, 170,
Also see Detrending

SUBJECT INDEX

— plus cycle model   64
Butterworth trend   56, 140
Fixed   60, 63, 65, 101
Global   67
Level   50, 55
    Fixed   47
    Stochastic   34, 53
Local level   50, 56, 140, 170–172, 178, 182
Local level with drift   56
Local linear trend   55, 173
Quadratic   87
Slope   50, 55
    Stochastic   34, 53
Smooth trend   56, 140, 177
Stochastic   56, 66, 77, 78, 80, 100, 169, 170
Trend-cycle decomposition   106
Tutorial data sets   14

Unobserved components   4, 5, 77, 140, 171, 175, 189, 203, 215

VAR   *see* Vector autoregression
Variable names in batch   215
Vector autoregression   6, 16, 92, 171, 190, 191, 194
    Stationarity   53
Volatility   *see* Stochastic volatility

Weight functions   5, 8, 107, 160
— graphics   40, 110, 131
White noise   52, 63, 106, 125, 209
Window   200
World Wide Web   13

X-11   *see* Bureau of the Census X-11